IDENTIFICATION GUIDE

WORLD WAR II TANKS: WESTERN ALLIES 1939 – 45

POLAND, BELGIUM, FRANCE, BRITAIN, CANADA, UNITED STATES

IDENTIFICATION GUIDE

WORLD WAR II TANKS: WESTERN ALLIES 1939 – 45

POLAND, BELGIUM, FRANCE, BRITAIN, CANADA, UNITED STATES

DAVID PORTER

This edition published in 2022

Reprinted in 2024

Published by
Amber Books Ltd
United House
London N7 9DP
United Kingdom
www.amberbooks.co.uk
Facebook: amberbooks
YouTube: amberbooksltd
Instagram: amberbooksltd
X(Twitter): @amberbooks

Copyright © 2009 Amber Books Ltd

All rights reserved. With the exception of quoting brief passages for the purpose of review, no part of this publication may be reproduced, stored in a retrieval system or transmitted in any form or by any means, electronic, mechanical, photocopying, recording or otherwise, without prior written permission from the publisher.

ISBN: 978-1-83886-113-1

Project Editor: Michael Spilling
Design: Hawes Design
Picture Research: Terry Forshaw

Printed in China

PHOTOGRAPH AND ILLUSTRATION CREDITS
Art-Tech/Aerospace: 25 (bottom), 69 (top), 87 (top), 95 (top), 102, 105 (top),
    106, 108 (top), 113 (top), 114 (bottom), 116, 120, 124 (bottom), 141 (top),
    152, 159, 161, 166 (top), 175
Art-Tech/MARS: 20
Cody Images: 6, 8, 9, 10, 15 (top), 16 (both), 22, 28 (top), 35, 44, 50, 52, 57 (top), 64,
    65 (top), 69 (bottom), 72, 75 (top), 76, 80 (top), 89 (top), 92, 99, 101 (bottom),
    114 (top), 122 (top), 128 (top), 129 (bottom), 133 (top), 149 (top), 150, 163 (top)
Public Domain: 12
U.S. National Archives: 136, 146 (top)

All artworks courtesy of Alcaniz Fresno's S.A. and Art-Tech/Aerospace

Contents

Introduction	6
Chapter 1 **Defence of Poland**	10
Chapter 2 **France and the Low Countries**	20
Chapter 3 **North Africa: 1940–43**	50
Chapter 4 **Sicily and Italy: 1943–45**	76
Chapter 5 **D-Day to Arnhem: 1944**	102
Chapter 6 **The Ardennes Offensive**	136
Chapter 7 **Invading the Reich**	152
Appendices	172
Bibliography	186
Index	187

Introduction

On 8 August 1918, a British force of 414 combat tanks and 120 supply tanks, supported by 800 aircraft, played a key role in the decisive Allied victory at Amiens. The impression was given that a true revolution in military technology had occurred that would change the face of warfare forever. The events of the next few months would show that Amiens had been a false dawn and that it would be another 20 years before technology caught up with the theories of visionaries such as J.F.C. Fuller, who devised the futuristic Plan 1919. The process of creating the tanks and other essential equipment that would transform the theory and practice of armoured warfare between 1939 and 1945 was to be long and hard.

◀ **Pre-war trials**
During the inter-war period, much research was carried out using the large stock of French Renault FT-17s. This vehicle is one of a number fitted with new tracks and suspension and designated the FT Kegresse-Hinstin M26/27.

INTRODUCTION

▲ **Elite French armour**
A column of H-35 light tanks of the *18e Dragons*, part of the elite 1re DLM.

IN 1918, IT SEEMED that a revolution in land warfare was imminent – British and French practical experience of armoured warfare was on the point of being linked to the industrial strength of the United States to produce a huge Allied mechanized force. However, this proved to be a false dawn – the unexpectedly swift collapse of the Central Powers drastically changed military priorities as the United States withdrew into isolationism, whilst the British and French armies largely reverted to their old 'colonial policing' roles.

Although US and French armoured forces soon came under infantry control, in Britain the Tank Corps just retained its independence. Most importantly, in 1923 it also managed to acquire a modern tank, the Vickers Medium. Whilst the design had many faults, it was able to serve as a test-bed for establishing which of the many theories of armoured warfare were truly practical. The most important trials were the exercises carried out in 1927–31 by the Experimental Mechanized Force, which pioneered the techniques of tanks, artillery, infantry and engineers all operating together under radio control.

Radio technology

The development of reliable radio was arguably one of the most significant technological advances affecting armoured warfare in the inter-war years, but improved weapons, engines, tracks and suspension systems were also vital in allowing theories to be transformed into reality. Possibly the greatest innovator of the period was an eccentric American inventor, J. Walter Christie, who devised a variety of amphibious and airmobile tanks.

Whilst these projects were asking too much of the technology of the time, his 'big wheel' suspension system, soon to be known as the Christie suspension,

INTRODUCTION

offered an immense improvement in cross-country performance and was adopted for a host of British and Soviet tank designs, including the Crusader, Cromwell and T-34.

French designers pioneered the use of very large castings for gun mantlets, turrets and eventually entire tank hulls, a technique also taken up by the US and USSR and, to a lesser extent, the UK. Most early tanks had used riveted or bolted armour, which was inherently dangerous as hostile fire could shear off the rivets and bolts, which became lethal projectiles flying around inside the vehicle. Welding overcame this vulnerability, although the welds had to be subjected to stringent quality control to ensure they could withstand hits from high-velocity weapons.

New weapons technology

The inter-war years were also marked by significant improvements in weapons technology as the machine guns and low-velocity guns of World War I gradually gave way to high-velocity weapons capable of destroying tanks at ranges out to 1000m (1095 yards) or more. Even so, the chances of acquiring and hitting a target at that sort of range remained minimal until well into the war.

KEY TO TACTICAL SYMBOLS USED IN ORGANIZATION CHARTS

Symbol	Description	Symbol	Description
	Symbol for division or larger	Sig	Signals unit
	Symbol for regiment or brigade-sized formation	Pio	Pioneer unit
		Sup	Support unit
	Symbol for battalion or smaller unit	Inf	Infantry unit
HQ	HQ units	Bat	Battery
Lt	Light tank unit (battalion or company)	Mnt	Maintenance unit
Med	Medium tank unit	Art	Artillery unit
Hv	Heavy tank unit	Mor	Mortar unit
AC	Armoured car unit	Rec	Reconnaissance
Mc	Motorcycle unit	AA	Anti-aircraft unit
Eng	Engineer unit	AG	Assault gun unit

▲ **Autotransport**
In 1931, a small number of tankette transporters, dubbed 'autotransports', were produced on modified Ursus truck chassis. The tankette was driven onto the autotransport, its tracks were removed and a chain drive was taken from the drive sprocket to the rear axle of the transporter. The tankette's driver was then able to control the combination from his usual position inside the tankette.

Chapter 1

Defence of Poland

The first Polish-crewed tanks were 150 Renault FT-17s that were delivered in 1919–20. In 1929, modified versions of Vickers-Carden-Loyd tankettes (TK and TKS) and Vickers 6-ton tanks (generally known as the Vickers E) were adopted after extensive trials. These formed the mainstay of the *Bron Pancerna*, or Armoured Forces, an organization set up in 1930 to administer all tanks, armoured cars and armoured trains. By 1935, a total of nine armoured battalions had formed, although these were actually deployed in company units (usually of 13 vehicles each) supporting infantry divisions and cavalry brigades.

◀ Training
A column of Polish 7TPs, based on the Vickers 6-Ton tank, on pre-war manoeuvres.

DEFENCE OF POLAND

On the road to war

The increasing threat from German and Soviet forces led to Poland's adoption of an ambitious six-year plan in 1936 under which a total of 479 20mm (0.79in) gun armed 4TP light tanks were to be built for infantry and cavalry support.

EIGHT BATTALIONS OF 7TPs (the latest version of the 6-Tonner) were to be allotted to army group commanders. The most radical feature of the plan was the formation of four Mechanized Brigades (OMs), each of which was to include two tank battalions: one of 7TPs and the other a mixed unit of 4TPs and new 10TP fast medium tanks, the latter model to have Christie suspension, 20mm (0.79in) armour and 37mm (1.5in) main armament. Each OM would also have two battalions of motorized infantry, and a battalion each of artillery, anti-tank guns and engineers.

Imported tanks

Poland could not pay for such an ambitious modernization programme and in 1936 France agreed to supply tanks and to make a loan to cover their cost. Even this did not solve the problem as French defence industries were working flat out to supply their own forces and little could be spared for export. An order for Somua S-35s had to be turned down, and although it was agreed to supply 100

Armoured Fighting Vehicles (AFVs), September 1939	Strength
TK and TKS Tankettes	440
7TPs	130
Vickers E	30
R-35s	49
Renault FT-17s	55
wz.29 and wz.34 Armoured Cars	95

R-35s instead, only 53 had been delivered before the German invasion. Besides these tanks and tankettes, almost 100 armoured cars were in service with the reconnaissance squadrons of the 11 cavalry brigades, generally deployed in troops of eight cars each. The approximate totals of operational Polish AFVs as at 1 September 1939 are set out in the table above.

The German *Panzerwaffe*

The evolution of major German armoured units began in 1934 with the formation of the first operational Panzer I battalion. The pace of events quickened throughout the 1930s as German confidence grew, especially after Hitler formally announced his intention to ignore military restrictions imposed on Germany by the 1919 Treaty of Versailles. By 1938, when Heinz Guderian, General Oswald Lutz's former chief of staff, was appointed *General der Panzertruppen*, the first four Panzer divisions had been formed.

All this effort could not alter the fact that Germany's resources were limited and that tank production had to compete with the needs of the growing *Luftwaffe* and *Kriegsmarine*. As a result, although the official strength of a Panzer division at the beginning of the war totalled 562 tanks, the reality was very different, with the strongest formations only fielding 328 tanks each.

Of these, the vast majority were Panzer Is and IIs, which even by 1939 standards were only really fit for training and reconnaissance. The Panzer IIIs and IVs – intended to be the main combat tanks – were in

▲ **Polish Vickers Mark E light tank**
An example of the Polish twin-turreted Vickers E in its initial configuration – a total of 38 were delivered. The tank is armed with two 7.92mm (0.31in) wz.25 Hotchkiss machine guns, which were later replaced with water-cooled wz.30 weapons in 16 vehicles. (Most of the remainder received a wz.30 in the left-hand turret and a 13.2mm (0.52in) Hotchkiss heavy machine gun in the right.)

DEFENCE OF POLAND

short supply and formed only a small proportion of the total armoured strength. It was only the reinforcement of the *Panzerwaffe* with the excellent Czech Panzer 35(t)s and 38(t)s which made it a truly battleworthy force. Most of these were distributed among the five Panzer divisions and four light divisions, which were assigned to the Polish operation.

Tank Formations, 1939	Sub-Unit	TK	TKS	7TP	R–35	Vickers	FT–17	wz.29	wz.34
Mazowska BK	11 dp	–	13	–	–	–	–	8	–
Wolynska BK	21 dp	–	13	–	–	–	–	–	8
Suwalska BK	31 dp	–	13	–	–	–	–	–	8
Podlaska BK	32 dp	–	13	–	–	–	–	–	8
Wilenska BK	33 dp	–	13	–	–	–	–	–	8
Krakowska BK	51 dp	13	–	–	–	–	–	–	8
Kresowa BK	61 dp	–	13	–	–	–	–	–	8
Podolska BK	62 dp	–	13	–	–	–	–	–	8
Wielkopolska BK	71 dp	–	13	–	–	–	–	–	8
Pomorska BK	81 dp	13	–	–	–	–	–	–	8
Nowogrodzka BK	91 dp	13	–	–	–	–	–	–	8
WBP–M	11 sk	–	13	–	–	–	–	–	–
WBP–M	12 sk	–	13	–	–	–	–	–	–
25 DP	31 sk	–	13	–	–	–	–	–	–
10 DP	32 sk	–	13	–	–	–	–	–	–
30 DP	41 sk	13	–	–	–	–	–	–	–
Kresowa BK	42 sk	13	–	–	–	–	–	–	–
GO Bielsko	51 sk	13	–	–	–	–	–	–	–
GO Slask	52 sk	13	–	–	–	–	–	–	–
GO Slask	61 sk	–	13	–	–	–	–	–	–
20 DP	62 sk	–	13	–	–	–	–	–	–
8 DP	63 sk	–	13	–	–	–	–	–	–
26 DP	71 sk	–	13	–	–	–	–	–	–
14 DP	72 sk	–	13	–	–	–	–	–	–
4 DP	81 sk	13	–	–	–	–	–	–	–
26 DP	82 sk	13	–	–	–	–	–	–	–
10 DP	91 sk	13	–	–	–	–	–	–	–
10 DP	92 sk	13	–	–	–	–	–	–	–
10 BKM	101 sk	13	–	–	–	–	–	–	–
10 BKM	121 sk	–	13	–	–	–	–	–	–
	1 bcl	–	–	49	–	–	–	–	–
	2 bcl	–	–	49	–	–	–	–	–
	21 bcl	–	–	–	45	–	–	–	–
	111, 112, 113 kcl	–	–	–	–	–	45	–	–
WBP–M	12 kcl	–	–	–	–	17	–	–	–
10 BKM	121 kcl	–	–	–	–	17	–	–	–

KEY
- BK — *Brygada kawalerii* — Cavalry brigade
- BKM — *Brygada kawalerii mechanizowanej* — Mechanized cavalry brigade
- DP — *Dywizja piechoty* — Infantry division
- dp — *Dywizjon pancerny* — Armoured troop
- GO — *Grupa operacjna* — Operational group
- kcl — *Kompanja czogow lekkich* — Light tank company
- sk — *Samodziena kompanja* — Scout tank company
- WBP-M — *Warszawska Brygada Pancerno-Motorowa* — Warsaw Mechanized Brigade

DEFENCE OF POLAND

First clash of armour
1 September 1939

Although relations with Germany had been strained for months, the invasion caught the Poles in the throes of mobilization, which had only been ordered on 30 August when signs of the coming attack were unmistakable.

THE GERMAN TAKEOVER of the Sudetenland in October 1938 followed by what was left of Czechoslovakia in March 1939, allowed offensives to be launched from an arc of territory stretching for over 1000km (625 miles) along Poland's northern, western and southern frontiers from East Prussia, through Pomerania and Silesia to Slovakia. The overall objective was to encircle the Polish armies in a great pincer movement, with the jaws of the pincer closing east of Warsaw, before destroying the trapped units in a classic *Kesselschlacht*, or cauldron battle.

Plan Z
In contrast, the Polish 'Plan Z', adopted in March 1939, intended to concentrate forces against each main German attack to slow its progress before gradually falling back on Warsaw, taking up successive delaying positions along the rivers of central Poland. It was estimated that this strategy could hold the Germans for up to six months, giving time for the French to launch their promised offensives against the weakly garrisoned Rhineland and Ruhr in the West.

A number of factors combined to wreck this plan, including German air superiority and the concentrated armoured punch provided by the six Panzer and four light divisions. Only one of the four planned OMs was ready for action and this, the 10th Mechanized Brigade, was well under strength. Although heavily outnumbered, this formation mauled 2nd Panzer and 4th Light Divisions as they advanced on Cracow, but it could not hope to win a decisive victory.

Polish armoured attack
Apart from the action near Cracow, the only major clash of armoured units came when the incomplete Warsaw Mechanized Brigade played a key role in defeating 4th Panzer Division's attempt to take the city on 9 September.

This attack was launched before the tired German infantry divisions (which remained dependent on horse-drawn transportation throughout the war) could catch up with the Panzers. As a result, the only infantry available were 4th Panzer's own two battalions, which were easily held by the city's garrison whilst the Warsaw Mechanized Brigade and anti-tank guns dealt with the German AFVs.

By the end of the day's fighting, 57 German tanks had been destroyed, 40 of them by a single platoon of 7TPs and a company of nine Bofors 37mm (1.5in) anti-tank guns. After this, attacks on the capital were left to infantry units with heavy air and artillery support until the garrison surrendered on 27 September.

Small-scale actions
A few small-scale armoured actions were fought after 4th Panzer's abortive attack on Warsaw. Pre-war plans had called for a high proportion of the TK and TKS tankettes to be rearmed with 20mm (0.79in) cannon, but only 20–25 TKSs had been converted by the time of the invasion in September 1939. A few cannon-armed tankettes saw action with the first platoon of the 71st Armoured Battalion of the *Wielkopolska* Cavalry Brigade.

On 14 September, its tankettes supported an attack by the brigade's 7th Mounted Rifles Regiment at Brochów (on the River Bzura, east of Warsaw). During the skirmish, a cannon-armed TKS fired across the river, knocking out two or three tanks of 4th Panzer Division. In a further action on 18 September near Pociecha in Puszcza Kampinoska forest east of Warsaw, three tankettes sprang an ambush at a forest crossroads.

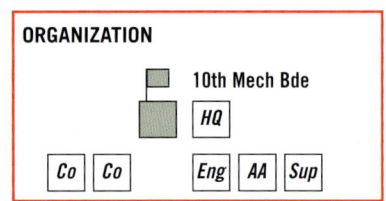
ORGANIZATION
10th Mech Bde
HQ
Co | Co | Eng | AA | Sup

DEFENCE OF POLAND

▲ **Awaiting the scrap heap**
Burnt-out Renault FT-17s are examined by German troops following the invasion of Poland, September 1939.

▶ **wz.34 Armoured Car**
11th Experimental Armoured Battalion, Modlin
As a result of mechanical problems, it was decided to rebuild the 90 wz.28 armoured halftracks in Polish service as wz.34 armoured cars. The leisurely conversion programme ran from 1934 to 1938. Roughly 60 vehicles were completed as shown, armed with a single Hotchkiss wz.25 machine gun.

 Most Polish vehicles were camouflaged in a three-colour scheme of light sand, olive green and a very dark chestnut brown. These colours were spray-painted in irregular horizontal patterns, with feathered edges. The main exception to this was the R-35 battalion, whose tanks were left in their original olive green. No Polish nationality markings were used and the few authorized peacetime unit insignia were removed on mobilization.

Specifications
Crew: 2
Weight: 2.2 tonnes (2.16 tons)
Length: 3.62m (11ft 11in)
Width: 1.91m (6ft 3in)
Height: 2.22m (7ft 3in)
Engine: 17.88kW (24hp) FIAT-108-III (PZIn".117) petrol
Speed: 50km/h (31mph)
Range: 250km (155 miles)
Armament: 1 x 7.92mm (0.31in) Hotchkiss wz.25 MG

DEFENCE OF POLAND

One of these was a cannon-armed vehicle, commanded by Cadet Roman Orlik, who destroyed a patrol of three Panzer 35(t)s of 11th Panzer Regiment, 1st Light Division, with hits on their side armour. On the following day, the tankette platoon supported the defenders of Sierakόw village in Puszcza Kampinoska against an attack by 11th Panzer Regiment. Orlik's TKS carried out short forays from tank scrapes on the left flank of the German advance and claimed the destruction of seven tanks. (Unconfirmed Polish reports claimed that the Germans lost a total of 20 tanks in the action, mostly due to anti-tank and field artillery.) A single surviving cannon-armed tankette of the 71st Armoured Battalion got through to Warsaw and took part in its defence, but no further details have emerged.

By this time, the Poles' last chance of holding out had been wrecked by the Soviet invasion of their country, which began on 17 September. Stalin quickly achieved his objective of securing eastern Poland, as had been agreed by the Russo-German Non-Aggression Pact signed in August 1939. Caught between the attacks from east and west, by the beginning of October the remnants of *Bron Pancerna* had been destroyed or had escaped across the frontier to internment in neutral Rumania.

▲ **Booty**
A captured TKS tankette. A number were taken into German service as the TKS(p).

▼ **Winter manoeuvres**
A TKS tankette tows a detachment of ski troops on pre-war winter manoeuvres.

DEFENCE OF POLAND

▶ wz.34 Armoured Car

Lodz Army / Wolynska Cavalry Brigade / 21st Armoured Battalion / Armoured Car Squadron

About 30 vehicles were fitted with an L/21 37mm (1.5in) SA-18 Puteaux gun instead of the Hotchkiss machine gun. These gun-armed cars were generally used by squadron and troop commanders.

Specifications
Crew: 2
Weight: 2.2 tonnes (2.16 tons)
Length: 3.62m (11ft 11in)
Width: 1.91m (6ft 3in)
Height: 2.22m (7ft 3in)
Engine: 17.88kW (24hp) FIAT-108-III (PZIn".117) petrol
Speed: 50km/h (31mph)
Range: 250km (155 miles)
Armament: 1 x 37mm (1.5in) wz.18 (SA-18) Puteaux gun

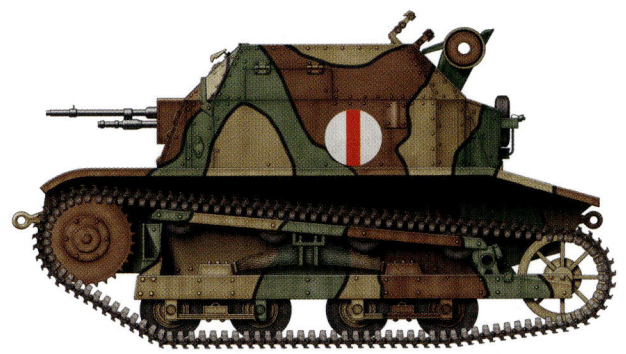

▲ TK Tankette

Cracow Army / Operational Group Bielsko / 51st Independent Reconnaissance Tank Company

The TK tankette was developed from the Carden-Loyd carrier series. A total of possibly 300 TK tankettes were built between 1931 and 1933.

Specifications
Crew: 2
Weight: 2.43 tonnes (2.39 tons)
Length: 2.58m (8ft 5in)
Width: 1.78m (5ft 10in)
Height: 1.32m (4ft 4in)
Engine: 29.8kW (40hp) Ford A 4-cylinder petrol
Speed: 46km/h (28.58mph)
Range: 200km (124 miles)
Armament: 1 x 7.92mm (0.31in) Hotchkiss wz.25 MG

▶ TKS Tankette

Poznan Army / Podolska Cavalry Brigade / 62nd Armoured Battalion / Reconnaissance Tank Squadron

The TKS closely resembled the TK tankette but incorporated a number of improvements, including thicker armour, a more powerful engine and wider tracks.

Specifications
Crew: 2
Weight: 2.65 tonnes (2.60 tons)
Length: 2.58m (8ft 5in)
Width: 1.78m (5ft 10in)
Height: 1.32m (4ft 4in)
Engine: 34.27kW (46hp) Polski FIAT-122BC 6-cylinder petrol
Speed: 40km/h (25mph)
Range: 180km (112 miles)
Armament: 1 x 7.92mm (0.31in) Hotchkiss wz.25 MG

DEFENCE OF POLAND

▶ 7TP Light Tank
Warsaw Defence HQ / Light Tank Company

A total of 24 twin-turreted 7TP light tanks were produced between 1935 and 1936. These vehicles were developed from the Vickers E, but were fitted with the new licence-built Saurer diesel engines and slightly thicker armour. These were always regarded as an interim type and were scheduled to be rebuilt as single-turreted vehicles, although not all had been converted by September 1939.

Specifications
Crew: 3
Weight: 9.4 tonnes (9.25 tons)
Length: 4.56m (14ft 11in)
Width: 2.43m (7ft 11in)
Height: 2.19m (7ft 2in)
Engine: 80kW (110hp) Saurer VBLDd diesel
Speed: 37km/h (23mph)
Range: 160km (99.42 miles)
Armament: 2 x 7.92mm (0.31in) Ckm wz.30 MGs (1 in each turret)

◀ 7TP Light Tank
Reserve Army 'Prussia' / 2nd Light Tank Battalion

At the time of the German invasion, the bulk of the operational 7TPs (98 vehicles) were serving with the 1st and 2nd Light Tank Battalions, both of which were assigned to Reserve Army 'Prussia'. The tank's L/45 37mm (1.5in) wz.37 gun (a licence-built copy of a Bofors design) proved highly effective against all German AFVs encountered during the campaign.

Specifications
Crew: 3
Weight: 9.9 tonnes (9.74 tons)
Length: 4.56m (14ft 11in)
Width: 2.43m (7ft 11in)
Height: 2.3m (7ft 6in)
Engine: 80kW (110hp) Saurer VBLDd diesel
Speed: 37km/h (23mph)
Range: 160km (99 miles)
Armament: 1 x 37mm (1.45in) Bofors wz.37 gun, plus 1 coaxial Ckm wz.30 MG

Specifications
Crew: 1
Weight: 1.85 tonnes (1.82 tons)
Length: 4.72m (15ft 6in)
Width: 2m (6ft 7in)
Height: 2.11m (6ft 11in)
Engine: 38.75kW (52hp) petrol
Speed: 60km/h (37mph)
Range: 400km (248 miles)

▲ Polski Fiat PF-618 Light Truck
Poznan Army / 26th Infantry Division / 82nd Independent Reconnaissance Tank Company / Maintenance Platoon

The PF-618 was a relatively modern light truck, which equipped the support units of many armoured formations.

DEFENCE OF POLAND

▼ Warsaw Mechanized Brigade

A second OM, the Warsaw Mechanized Brigade, was forming at the time of the German invasion. It had one 7TP company (16 vehicles — one command tank plus three platoons each of five tanks) and two TKS companies (13 vehicles each — one command tankette plus a supply platoon of two vehicles and two platoons with five tankettes each). Most of the remaining Polish tank units fought as battalions — the 1st and 2nd Light Tank Battalions each had 49 7TPs, whilst the 21st was equipped with 45 R-35s.

Company Command

1 Platoon

2 Platoon

3 Platoon

Company Command    **Company Supply**

1 Platoon

2 Platoon

Company Command    **Company Supply**

1 Platoon

2 Platoon

Chapter 2

France and the Low Countries

In 1939, the prestige of the French Army was immense. Its performance on the battlefields of World War I had won the respect of allies and enemies alike. During the inter-war years it had constructed the impressive fortifications of the Maginot Line and developed some of the most formidable tanks in the world. Despite this show of power, it was crippled by weaknesses, especially in its command structure, that were only apparent to a small minority of observers. As the Germans prepared to launch *Fall Gelb* (Case Yellow), the great offensive in Western Europe, everything depended on how far they could exploit those failings in their opponents.

◀ **Alpine manoeuvres**
Char Léger R-35 tanks move through an Alpine pass on training manoeuvres, August 1938.

FRANCE & THE LOW COUNTRIES

Netherlands and Belgium

As World War II began, the Netherlands had not fought a war since the Belgian Revolution of 1830. The country had a very small military industrial base and its army was in a sorry state after the economic crises of the inter-war years.

THE BELGIAN ARMY had used armoured cars as early as 1914, but the economic crises of the 1920s and 1930s, coupled with a foreign policy of strict neutrality, had prevented the development of any significant armoured forces. Whilst almost 300 AFVs were in service at the time of the German invasion in May 1940, they were parcelled out in 'penny packets' to various infantry and cavalry divisions for use as support weapons.

Poor relation

The Dutch Army was in an even worse position than its Belgian counterpart – it had been subjected to massive budget cuts before World War II, and at the time of the German invasion of the Netherlands its only AFVs were 41 armoured cars and five Carden-Loyd machine-gun carriers.

The Great Depression had hit the Dutch Army especially hard – conscripts' terms of service were progressively cut from two years to six months, barely sufficient for basic training. As early as 1925, a report indicated that the army required 350 million guilders for modernization, but all that happened was a further cut of 100 million guilders. A Dutch parliamentary committee set up to investigate the possibility of further cuts in the military budget reported that the army was in such a poor state that any more cutbacks would seriously endanger its sustainability. The committee was dissolved and

Belgian Armoured Distribution, May 1940	T.13	T.15	ACG-1
1re Division d'Infanterie	12	–	–
2e Division d'Infanterie	12	–	–
3e Division d'Infanterie	12	–	–
4e Division d'Infanterie	12	–	–
7e Division d'Infanterie (R)	12	–	–
8e Division d'Infanterie (R)	12	–	–
9e Division d'Infanterie (R)	12	–	–
10e Division d'Infanterie (R)	12	–	–
11e Division d'Infanterie (R)	12	–	–
1re Division des Chasseurs Ardennais	48	3	–
2e Division des Chasseurs Ardennais	–	3	–
Compagnie Ind. de l'Unité Cyclistes Frontière	12	–	–
8e Companie de l'Unité Cyclistes Frontière	12	–	–
Compagnie PFL de l'Unité de Forteresse	12	–	–
1re Division de Cavalerie			
1er Régiment de Guides	6	6	–
2e Régiment de Lanciers	6	6	–
3e Régiment de Lanciers	6	6	–
2e Division de Cavalerie			
1er Régiment de Lanciers	6	6	–
1er Régiment de Chasseurs	6	6	–
2e Régiment de Chasseurs	6	6	–
Escadron d'Autoblindées	–	–	8

▲ **Phoney War**
A Belgian T.15 light tank and motorcycle despatch rider wait by the roadside during the Phoney War period.

replaced by a new body that was far more ruthless and managed to make cuts totalling an additional 160 million guilders.

Although funding was reluctantly made available for some modernization in the late 1930s, its effects were limited. A total of 400 modern Bohler 47mm (1.9in) anti-tank guns were bought, but much of the army's light artillery was archaic – over 100 of its field guns dated back to 1878 and a further 200 were of 1894 vintage. As late as February 1940, the defence minister refused to release funding to modernize all the Dutch defence lines, provoking the resignation of the commander-in-chief.

In the circumstances, it is hardly surprising that Dutch armoured forces were minimal. Nonetheless, the traditional Dutch expertise in the use of water obstacles was still highly effective, and the handful of armoured cars that did manage to get into action against German paratroops indicated the value of even light armour against airborne forces.

▶ M39 Pantserwagen
3e Eskadron Pantserwagens
The M39 was a DAF-designed armoured car that was just entering Dutch service at the time of the German invasion.

Specifications
Crew: 6
Weight: 6.9 tonnes (6.79 tons)
Length: 4.63m (15ft 2in)
Width: 2m (6ft 7in)
Height: 2m (6ft 7in)
Engine: 70.77kW (95hp) Ford 8-cylinder petrol
Speed: 60km/h (37mph)
Range: 300km (180 miles)
Armament: 1 x 37mm (1.5in) Bofors gun, plus 3 x 7.92mm (0.31in) MGs (1 coaxial, 1 ball-mounted in hull front and 1 ball-mounted in hull rear)

◀ M39 Pantserwagen
3e Eskadron Pantserwagens
It seems likely that no more than a dozen or so M39s were completed, of which one or two may have seen action alongside the earlier M36 and M38 armoured cars in 1940.

Specifications
Crew: 6
Weight: 6.9 tonnes (6.79 tons)
Length: 4.63m (15ft 2in)
Width: 2m (6ft 7in)
Height: 2m (6ft 7in)
Engine: 70.77kW (95hp) Ford 8-cylinder petrol
Speed: 60km/h (37mph)
Range: 300km (180 miles)
Armament: 1 x 37mm (1.5in) Bofors gun, plus 3 x 7.92mm (0.31in) MGs (1 coaxial, 1 ball-mounted in hull front and 1 ball-mounted in hull rear)

▶ Vickers Utility Tractor
1re Division d'Infanterie
This armoured version of the Vickers one-man utility tractor was widely used by the Belgian Army for towing anti-tank guns and for front-line supply duties.

Specifications
Crew: 1
Weight: not known (n/k)
Length: 2.13m (7ft)
Width: 1.22m (4ft)
Height: 1.52m (5ft) (estimated)
Engine: 38.74kW (52hp) Ford 4-cylinder petrol
Speed: 16km/h (10 mph) (estimated)
Range: n/k

FRANCE & THE LOW COUNTRIES

Belgium, meanwhile, had adopted, in 1936, a policy of 'armed independence', rather than a stance of strict neutrality, entailing an increase in defence expenditure to almost 25 per cent of the national budget. Belgian attitudes towards both Britain and France were ambivalent – both countries had been allies during World War I, but post-war disagreements had soured relations. This was especially true in the case of France – there was a popular perception in Belgium that French generals and politicians saw Belgium not as an independent state but as an ideal battlefield for any future war with Germany. This perception was strengthened by the failure to extend the Maginot Line to the Channel coast – to Belgian public opinion, this was tantamount to a French invitation to the Germans to attack France through Belgium. The French government turned down several requests to extend the fortifications, which increased Belgian suspicions of French intentions.

Belgian armour

By 1940, the Belgian Army totalled 650,000 men in 22 divisions. It lacked modern AA guns, but some of its equipment was good, such as the 750-plus 47mm (1.9in) FRC Model 1931 anti-tank guns, which were issued on a scale of 32 per division. Each division also had a company of 12 T.13 tank destroyers, light AFVs mounting the same 47mm (1.9in) gun.

In all, the Belgian Army fielded almost 300 AFVs at the time of the German invasion, but their effectiveness was greatly reduced by being spread thinly among infantry and cavalry divisions as mentioned above. The only exception to this type of deployment was the *1re Division des Chasseurs Ardennais*, which had 48 T.13s.

◀ **Renault ACG-1 Light Tank**
Escadron d'Autoblindées
The ACG-1 was the most potent AFV in the Belgian inventory in 1940, but only eight vehicles were operational at the time of the German invasion.

Specifications

Crew: 3	Engine: 134.1kW (180hp) Renault water-cooled 4-cylinder petrol
Weight: 14.5 tonnes (14.27 tons)	
Length: 4.57m (15ft)	Speed: 40km/h (24.85mph)
Width: 2.23m (7ft 3in)	Range: 161km (100 miles)
Height: 2.33m (7ft 8in)	Armament: 1 x 47mm (1.85in) SA35 L/32 gun, plus 1 x coaxial 13.2mm (0.52in) HMG

▲ **Ford/Marmon-Herrington Armoured Car**
2e Division de Cavalerie / 1er Régiment de Chasseurs
A total of 90 of these vehicles were produced. The chassis were built by Ford at Antwerp and the armoured bodies by Rageno at Malines. They were issued to cavalry regiments as prime movers for the 47mm (1.9in) anti-tank gun.

Specifications

Crew: 2	Engine: Ford V8 cylinder 63.38kW (85hp) petrol
Weight: 3.49 tons (7710lbs)	Speed: n/k
Length: 4.82m (15ft 9in)	Range: n/k
Width: 1.9m (6ft 3in)	Radio: none
Height: 1.77m (5ft 9in)	

FRANCE & THE LOW COUNTRIES

▶ T.15 Light Tank
1re Division de Cavalerie / 3e Régiment de Lanciers

The T.15 was another Vickers design developed specifically to meet Belgian requirements, with a tall conical turret mounting a Hotchkiss 13.2mm (0.52in) heavy machine gun.

Specifications

Crew: 2	Engine: 67kW (90hp) Meadows petrol
Weight: 6 tonnes (5.9 tons)	Speed: 62km/h (38.5mph)
Length: 3.4m (11ft 2in)	Range: n/k
Width: n/k	Armament: 1 x 13.2mm (0.52in) Hotchkiss HMG
Height: n/k	

The evolution of French armour
1918–40

By 1918, French tank design and production had reached an advanced stage under the energetic guidance of General Jean-Baptiste Estienne, the commander of all armoured units, which were termed *Artillerie d'Assaut*.

THEIR MOST IMPORTANT equipment was the Renault FT, the first turreted tank of modern layout to be produced in quantity, with a total of over 3000 delivered by November 1918. On occasions these had been used en masse to considerable effect, most notably near Soissons on 18 July 1918, where 225 FTs played the key role in the success of a major French counterattack that captured 10,000 prisoners and 200 guns in a single day.

After the end of World War I, General Estienne remained in charge of the technical development of French AFVs and his influence led to an army directive of July 1920 calling for a full range of vehicles to include:
- Light tanks
- Howitzer-armed support tanks
- *Chars de Rupture* – heavy breakthrough tanks. (Prototypes of these had been tested by 1918 and the definitive 69-tonne/68-ton Char 2C was ready for production. An order for 300 vehicles was placed in February 1918, but this was cut back to 10, all of which were delivered by 1922)
- Armoured supply and command vehicles.

All this was far too radical for the French military establishment, and the headquarters of the *Artillerie d'Assaut* was disbanded in 1920, whilst in 1921 a special commission ordered that the earlier directive should be cancelled. In future, tanks were to be committed to infantry support and were to come under infantry control.

Stalled development
This move stifled any major development of French armour during the 1920s and early 1930s when the

▲ **On parade**
H-35 light tanks on parade during the Phoney War period, late 1939.

FRANCE & THE LOW COUNTRIES

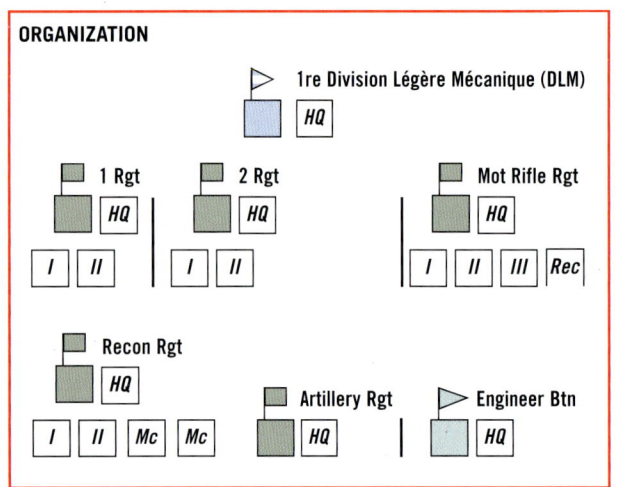

1RE DIVISION LEGERE MECANIQUE (1935)	Strength
Regiments x 2	
H-35s (per rgt)	40
S-35s (per rgt)	40
Recon Rgt with P-178s	40
Motorcycle Squadrons x 2	
Motorized Rifle Regiment (3 battalions)	
AMR-33/AMR-35s	60
Artillery Regiment	
75mm (2.9in) Field Guns	24
105mm (4.1in) Howitzers	12
Engineer Battalion x 1	

planning and construction of the Maginot Line claimed the lion's share of the limited defence budget.

Largely as a result, any real progress had to wait until 1934 with the formation of the *1re Division Légère Mécanique* (DLM). This was a mechanized cavalry division trained to operate in the classic cavalry roles of screening, strategic reconnaissance and pursuit. At first, there was very little modern equipment but gradually a range of new AFVs came into service, including the Somua S-35, the Renault R-35 and the Hotchkiss H-35.

Independent tank battalions

The impressive Char B, constructed by several producers, including Renault, entered service in 1935–36 with independent tank battalions and these were (in theory) combined with units of older medium tanks such as the Char D. All such tanks were officially described as *Chars de manoeuvre d'ensemble* ('operating together') and were tasked with the breakthrough role, but it was not until September 1939 that they were brought together in a larger permanent formation. This was the *1re Division Cuirassée de Réserve* (DCR).

The French tank industry could not produce enough Char B vehicles to equip further divisions on this scale and, to allow the 2e DCR to form by January 1940, it was decided to standardize the tank strengths of all DCRs as two battalions of Char Bs and two battalions of H-39s, each with 45 tanks. The other elements of the division remained unchanged.

The 3e DCR assembled in March 1940 and the 4e DCR, commanded by General Charles de Gaulle, was still incomplete when it was rushed into action

▶ **Renault Chenillette UE, experimental armed prototype**
In November 1931, the *Section Technique de la Cavalerie* asked Renault to rebuild one of its six *chenillette* prototypes as an armed tankette. Accordingly, Prototype No. 77982 was converted to an *Automitrailleuse légère de contact tout terrain*. A ball-mounted machine gun was installed in the front plate of a small rectangular superstructure fitted over the commander's position, but the type was rejected for service due to its low speed.

Specifications
Crew: 2
Weight: 2.64 tonnes (2.6 tons)
Length: 2.8m (9ft 2in)
Width: 1.74m (5ft 8in)
Height: 1.25m (4ft 1in)
Engine: 28.31kW (38hp) Renault 85 4-cylinder petrol
Speed: 30km/h (19mph)
Range: 100 km (62 miles)
Armament: 1 x 7.5mm (0.295in) MG in hull front (prototype only)

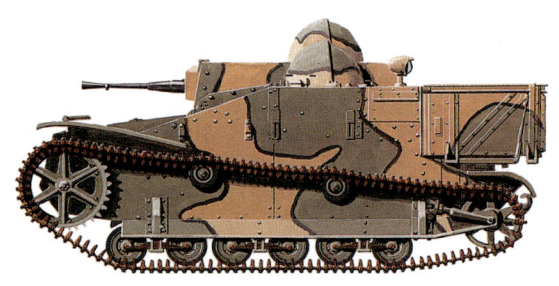

FRANCE & THE LOW COUNTRIES

▲ **Char de Cavalerie 38H (H-39)**

3e DLM / 1e Cuirassiers

The H-39 was a modernized H-35, fitted with a more powerful (90kW/120hp) engine, together with improved suspension and tracks. From April 1940, the L/35 37mm (1.5in) SA-38 gun was fitted to all new vehicles, with older H-39s being refitted with the weapon as supplies became available.

Specifications
Crew: 2
Weight: 12.1 tonnes (11.9 tons)
Length: 4.23m (13ft 10in)
Width: 1.96m (6ft 5in)
Height: 2.16m (7ft 1in)
Engine: 90W (120hp) Hotchkiss 1938 6-cylinder petrol
Speed: 36.5km/h (22.7mph)
Range: 150km (93 miles)
Armament: 1 x 37mm (1.5in) SA-38 gun, plus 1 x coaxial 7.5mm (0.295in) MG

◀ **Char de Cavalerie 35H (H-35)**

13e Bataillon de Chars de Combat (BCC)

Roughly 400 H-35s were in service in September 1939. It was intended to update them with improved vision devices and the L/35 37mm (1.5in) SA-38 gun with a much improved anti-tank capacity. The programme was never completed and many tanks went into action in their original configuration.

Specifications
Crew: 2
Weight: 10.6 tonnes (10.43 tons)
Length: 4.22m (13ft 10in)
Width: 1.96m (6ft 5in)
Height: 2.62m (8ft 7in)
Engine: 55.91kW (75hp) Hotchkiss 1935, 6-cylinder petrol
Speed: 27.4km/h (17mph)
Range: 150km (93 miles)
Armament: 1 x 37mm (1.5in) SA-18 gun, plus 1 x coaxial 7.5mm (0.295in) MG

▶ **Char Léger 35R (R-35)**

4e DCR / 24e Bataillon de Chars de Combat

The R-35 was a 1934 design intended to replace the venerable Renault FT-17 as the French Army's standard infantry support tank. Production began in 1936, but technical difficulties with the turrets delayed deliveries and the FT-17 had to remain in service for far longer than expected. Roughly 900 R-35s equipped front-line units by May 1940.

Specifications
Crew: 2
Weight: 14.5 tonnes (14.3 tons)
Length: 4.55m (14ft 11in)
Width: 2.2m (7ft 2.5in)
Height: 2.3m (7ft 6.5in)
Engine: 134.2kW (180hp) Renault 4-cylinder
Speed: 42km/h (26mph)
Range: 160km (99 miles)
Armament: 1 x 37mm (1.5in) SA-18 gun, plus 1 x coaxial 7.5mm (0.295in) MG

FRANCE & THE LOW COUNTRIES

▲ AMR-33
AMR-33 light tanks taking part in a pre-war Bastille Day parade.

following the German invasion in May 1940. In addition to these divisions, a total of five partially mechanized *Divisions Légères de Cavalerie* (DLCs) had formed during the winter and spring of 1939/40. These were little more than smaller and weaker versions of the DLMs. A total of over 3400 French tanks were available to oppose the German offensive that began in May 1940 – almost 2900 of these were modern types, with the balance being made up with old FTs and a few Char 2Cs. Their battlefield effectiveness was severely hampered by technical problems (especially short range, a lack of radios and, in most cases, one-man turrets). The deployment of

▶ Char Léger 35R (R-35)
4e DCR, 24e Bataillon de Chars de Combat

By the standards of 1940, the R-35 was an exceptionally well-protected light tank, with much of the vehicle covered with 40mm (1.57in) of armour, giving a high degree of immunity to contemporary German tank and anti-tank guns.

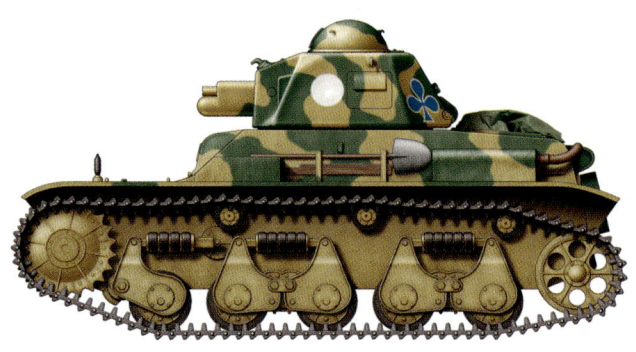

Specifications
Crew: 2
Weight: 14.5 tonnes (14.3 tons)
Length: 4.55m (14ft 11in)
Width: 2.2m (7ft 2.5in)
Height: 2.3m (7ft 6.5in)
Engine: 134.2kW (180hp) Renault 4-cylinder petrol
Speed: 42 km/h (26.1mph)
Range: 160km (99.4 miles)
Armament: 1 x 37mm (1.5in) SA-18 gun, plus 1 x coaxial 7.5mm (0.295in) MG

FRANCE & THE LOW COUNTRIES

nearly half the total strength in small units each of less than 50 vehicles scattered across the front under infantry command created added difficulties, making it impossible for the French to exploit their numerical superiority.

German AFV developments

Analysis of the victory in Poland had shown that the *Panzerwaffe* was working along the right lines, but that improvements were needed. The Panzer I was clearly in need of replacement, whilst the Panzer II was also obsolescent. As a result, every effort was made to re-equip units running these tanks with Panzer IIIs or Panzer 35(t)s and 38(t)s which had demonstrated their superiority over both these earlier types in the Polish campaign.

Work to meet the requirement for more specialized AFVs to support the main Panzer striking force was beginning to show results, with the first halftrack Armoured Personnel Carriers (APCs) coming into service along with small numbers of Stug III assault guns. A few heavier self-propelled (SP) guns and *Panzerjägers* ('Panzer hunters'), both types initially using obsolete Panzer I hulls, were also issued during the winter of 1939/40. The *Wehrmacht*'s total first line tank strength for the campaign against France

▶ **Char Léger 35R (R-35)**
21e Bataillon de Chars de Combat
The greatest weakness of most French tanks lay in their cramped one-man turrets, with a grossly overworked commander trying to act as loader and gunner as well as commanding the vehicle.

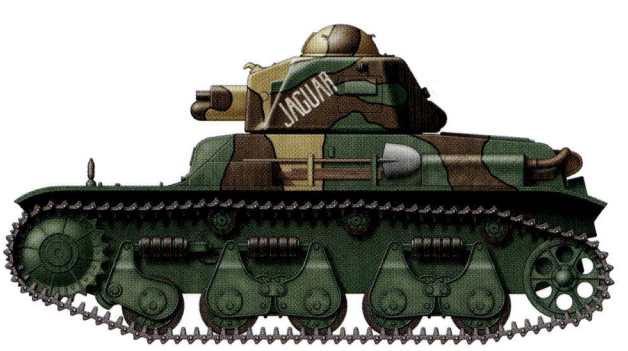

Specifications
Crew: 2
Weight: 14.5 tonnes (14.3 tons)
Length: 4.55m (14ft 11in)
Width: 2.2m (7ft 2.5in)
Height: 2.3m (7ft 6.5in)
Engine: 134.2kW (180hp) Renault 4-cylinder petrol
Speed: 42km/h (26mph)
Range: 160km (99 miles)
Armament: 1 x 37mm (1.5in) SA-18 gun, plus 1 x coaxial 7.5mm (0.295in) MG

Specifications
Crew: 3
Weight: 19.5 tonnes (19.2 tons)
Length: 5.38m (17ft 7.8in)
Width: 2.12m (6ft 11.5in)
Height: 2.62m (8ft 7in)
Engine: 141.7kW (190hp) V8 petrol
Speed: 41km/h (25mph)
Range: 257km (160 miles)
Armament: 1 x 47mm (1.9in) SA-35 gun, plus 1 x coaxial 7.5mm (0.295in) MG

▲ **Char de Cavalerie 35S (Somua)**
3e DLM / 2e Cuirassiers
The Somua was one of the best tanks of its time – with up to 47mm (1.9in) of armour and the potent L/32 47mm (1.9in) SA-35 gun, it was a formidable opponent for any German AFV.

FRANCE & THE LOW COUNTRIES

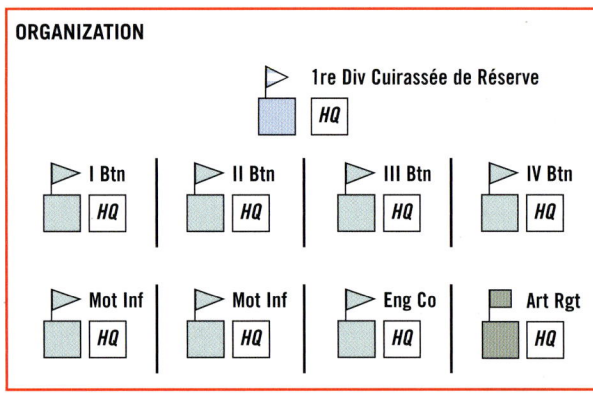

ORGANIZATION

stood at 640 Panzer Is, 825 Panzer IIs, 456 Panzer IIIs, 366 Panzer IVs, 151 Panzer 35(t)s and 264 Panzer 38(t)s.

In addition, there were nine batteries of six 150mm (5.9in) SP infantry guns as well as four independent Stug III batteries, with six assault guns in each. The infantry divisions also had the support of five *Panzerjäger* companies (two with 18 vehicles each, the remainder each having 27 vehicles).

Finally, there was a single heavy anti-tank company of 10 self-propelled 88mm (3.5in) Flak 18s mounted on armoured Sd Kfz 8 halftracks.

▲ **Char B1 bis**
1re DCR / 37e Bataillon de Chars de Combat
The Char B1 bis was a much-improved development of the earlier Char B. Orders were placed in 1937 for a total of 1144 vehicles, but only 369 examples of the B1 bis had been delivered by June 1940.

Specifications
Crew: 4
Weight: 32.5 tonnes (32 tons)
Length: 6.52m (21ft 5in)
Width: 2.5m (8ft 2in)
Height: 2.79m (9ft 2in)
Engine: 223.7kW (300hp) Renault 6-cylinder petrol
Speed: 28km/h (17mph)
Range: 180km (112 miles)
Armament: 1 x 75mm (2.9in) ABS SA-35 howitzer in the hull, 1 x 47mm (1.9in) SA-35 gun in turret, plus 2 x 7.5mm (0.295in) MGs (1 coaxial, 1 fixed forward-firing in hull)
Radio: ER51 or ER53

▶ **Panhard 178 Armoured Car**
6e Groupe de Reconnaissance de Division d'Infanterie (GRDI)
The Panhard 178 entered service in 1937, and 218 vehicles had been completed by September 1939. The three DLMs each had 40 vehicles, whilst each of the seven GRDIs fielded 12.

Specifications
Crew: 3
Weight: 8.63 tones (8.5 tons)
Length: 4.79m (15ft 8in)
Width: 2.01m (6ft 7in)
Height: 2.31m (7ft 7in)
Engine: 78KW (105hp) Panhard SK 4-cylinder petrol
Speed: 72 km/h (45mph)
Range: 300km (186 miles)
Armament: 1 x 25mm (0.98in) gun, plus 1 x coaxial 7.5mm (0.295in) MG
Radio: ER 26 or ER29

FRANCE & THE LOW COUNTRIES

▼ 1re Division Cuirassée de Réserve

The 33-strong Char B battalions of the DCRs were formidable assault forces, but required careful tactical handling due to the tanks' low speed. They were particularly vulnerable during the frequent refuelling halts imposed by their short range – 1re DCR was badly mauled by 7th Panzer Division whilst refuelling near Flavion on 15 May 1940.

Battalion

The battle for France
MAY–JUNE 1940

'La France a perdu une bataille, mais la France n'a pas perdu la guerre.' ('France has lost a battle, but France has not lost the war.') – General de Gaulle, BBC broadcast, 18 June 1940

THE MAGINOT LINE had an enormous impact on military planning in Europe throughout the 1930s. The popular press ran colourful articles often giving greatly exaggerated accounts of the strength of these fortifications, which covered the key areas of the French frontiers with Germany and Italy. To a large extent, the French fell victim to their own propaganda and came to rely too heavily on such fixed defences, whilst German planners naturally sought ways to avoid them. Until the beginning of 1940, the Germans were convinced that the only realistic option for a major attack on France was an offensive sweeping through Holland and Belgium to outflank the Maginot Line.

Standby

Most Allied experts agreed with this evaluation and from the beginning of the conflict, the British Expeditionary Force (BEF) and the strongest French field armies were on standby to move into Belgium and Holland as soon as the Germans attacked. The scene was set for stalemate, with both sides' most

FRANCE & THE LOW COUNTRIES

powerful forces meeting head-on in long, indecisive battles until an accident began a series of events leading to a stunning German victory.

On 10 January 1940, top secret documents outlining the plans for the German offensive were retrieved from a *Luftwaffe* communications aircraft that force-landed in Belgium after the pilot lost his way in bad weather. These papers were passed to French military intelligence and were welcomed as confirmation of their assessment of the situation. As a result, this compromised plan was replaced by General Erich von Manstein's radical proposal for a feint attack along the lines of the original plan by Army Group B (including three Panzer divisions) to draw Allied armour into Belgium.

Panzer thrust

The main offensive by Army Group A, spearheaded by seven Panzer divisions, would then strike through the weakly held wooded hills of the Ardennes and drive for the Channel coast. The Allied armies which had been drawn into Belgium would then be cut off and destroyed, virtually guaranteeing the final defeat of their remaining forces.

The German operation, designated *Fall Gelb* ('Case Yellow'), began on 10 May 1940 as airborne units landed at key points in Holland and Belgium whilst Army Group B advanced to link up with them. (The airborne landings in Holland suffered particularly heavy casualties, in part from the intervention of a handful of Dutch armoured cars.) The Allies reacted as planned, advancing to take up positions along the line of the River Dyle to meet the anticipated attack. As they moved up, the main German striking force was slowly pushing through the Ardennes, brushing aside light Belgian and French screening forces.

First encounter

The first major armoured action of the campaign took place in Belgium near the Gembloux Gap between the Rivers Meuse and Dyle, where the 2e and 3e DLMs clashed with the 3rd and 4th Panzer Divisions, which were significantly weaker than those deployed further south. Each of these Panzer divisions fielded 140 Panzer Is, 110 Panzer IIs, 150 Panzer IIIs and 40 Panzer IVs.

These were markedly inferior to the DLMs, each of which had 80 Somua S-35s plus 80 Hotchkiss H-35s or H-39s. The S-35s were particularly formidable as their 47mm (1.9in) guns could penetrate all the Panzers at normal battle ranges whilst they were immune to the German 37mm (1.5in) and short-barrelled 75mm (2.9in) guns except at point-blank range. It was only poor French battlefield tactics plus the slow, inaccurate fire of their one-man turrets that allowed the Panzers to manoeuvre and close the range to pick out their opponents' weak spots.

At the end of the action, each side had lost about 100 AFVs, but the DLMs withdrew to the main French positions, abandoning many damaged but repairable tanks. In contrast, the Panzer divisions were able to salvage most of their losses to help maintain their combat strengths.

As this two-day action ended on 13 May, the main German offensive developed along the Meuse. Six

---

INSIGNIA

French camouflage and markings were the least standardized and the most colourful of the period. Some vehicles, especially armoured cars, were a plain overall olive green (top), whilst others were over-painted with irregular disruptive patterns of ochre and/or dark chestnut brown (centre), often outlined in black.

After the opening of the German offensive, many AFVs had a varying number of national roundels painted on turret and hull tops and sides. (These roundels were sometimes also applied to the rear of the turret.)

INSIGNIA

The other widely used tactical markings were playing card symbols (indicating the company) superimposed on white shapes (defining the section). This system worked as follows:

Circle	1st Company (see top)
Square	2nd Company
Triangle	3rd Company
Spade*	1st Section (see centre)
Heart*	2nd Section
Diamond*	3rd Section
Club*	4th Section

*Blue for the 1st battalion and red for the 2nd battalion.

FRANCE & THE LOW COUNTRIES

hours of intense *Luftwaffe* attacks cracked open the defences that were largely held by low-grade French reserve infantry divisions, allowing bridgeheads to be established at Dinant, Montherme and Sedan. After the failure of the first counterattacks launched against these bridgeheads with local reserves, a larger-scale operation using all three DCRs was planned for 14 May. The lack of a unified command structure ruled out a prompt, properly coordinated attack, allowing each DCR to be defeated in detail.

1re DCR was destroyed as an effective fighting force after being caught refuelling by the 5th and 7th

AFV Units, 10 May 1940	R-35	H-39	D2	B1	FT-17
1re Division Cuirassée					
25e BCC	–	45	–	–	–
26e BCC	–	45	–	–	–
28e BCC	–	–	–	35	–
37e BCC	–	–	–	33	–
2e Division Cuirassée					
8e BCC	–	–	–	35	–
14e BCC	–	45	–	–	–
15e BCC	–	–	–	35	–
27e BCC	–	45	–	–	–
3e Division Cuirassée					
41e BCC	–	–	–	35	–
42e BCC	–	45	–	–	–
45e BCC	–	45	–	–	–
49e BCC	–	–	–	35	–
4e Division Cuirassée					
2e BCC	45	–	–	–	–
19e BCC	–	–	45	–	–
24e BCC	45	–	–	–	–
44e BCC	45	–	–	–	–
46e BCC	–	–	–	25	–
47e BCC	–	–	–	25	–
Compagnies Autonomes de Chars					
342e Cie A	–	15	–	–	–
343e Cie A	–	–	–	–	10
344e Cie A	–	–	–	–	10
345e Cie A	–	–	15	–	–
346e Cie A	–	–	15	–	–
347e Cie A	–	–	–	11	–
348e Cie A	–	–	–	11	–
349e Cie A	–	–	–	11	–
350e Cie A	–	–	–	–	10
351e Cie A	–	15	–	–	–
352e Cie A	–	–	–	11	–
353e Cie A	–	–	–	11	–

AFV Units, 10 May 1940	R-35	H-35	FCM-36	2C	FT-17
Bataillons Organiques					
1er BCC	45	–	–	–	–
3e BCC	45	–	–	–	–
4e BCC	–	–	45	–	–
5e BCC	45	–	–	–	–
6e BCC	45	–	–	–	–
7e BCC	–	–	45	–	–
9e BCC	45	–	–	–	–
10e BCC	45	–	–	–	–
11e BCC	–	–	–	–	63
12e BCC	45	–	–	–	–
13e BCC	–	45	–	–	–
16e BCC	45	–	–	–	–
17e BCC	45	–	–	–	–
18e BCC	–	–	–	–	63
20e BCC	45	–	–	–	–
21e BCC	45	–	–	–	–
22e BCC	45	–	–	–	–
23e BCC	45	–	–	–	–
29e BCC	–	–	–	–	63
30e BCC	–	–	–	–	63
31e BCC	–	–	–	–	63
32e BCC	45	–	–	–	–
33e BCC	–	–	–	–	63
34e BCC	45	–	–	–	–
35e BCC	45	–	–	–	–
36e BCC	–	–	–	–	63
38e BCC	–	45	–	–	–
39e BCC	45	–	–	–	–
40e BCC	45	–	–	–	–
43e BCC	45	–	–	–	–
48e BCC	45	–	–	–	–
51e BCC	–	–	–	6	–
Bataillon Colonial	–	–	–	–	63

FRANCE & THE LOW COUNTRIES

Panzer Divisions as they broke out from the Dinant bridgehead. XLI Panzer Corps (6th and 8th Panzer Divisions) struck 2e DCR as it attempted to deploy against the Montherme bridgehead, overrunning part of its artillery and scattering its transport. The DCR's armour fought on in battalions and smaller groups under local control, but it had lost the ability to inflict decisive damage on the Panzers.

Formidable opponent

Misguided orders rather than enemy action caused the destruction of 3e DCR, which was ready to attack the Sedan bridgehead when its tanks were ordered to be dispersed to form a 12.8km (8-mile) chain of mobile pillboxes along the front-line. This thin screen was easily broken when Guderian's XIX Panzer Corps (consisting of 1st, 2nd and 10th Panzer Divisions) advanced from Sedan. In these actions against the French DCRs, the Germans found that the Char B

Cavalry units, 10 May 1940	H-35	H-39	S-35	P178	AMR
1re Div Légère Mécanique					
4e Cuirassiers	40	–	40	–	–
18e Dragons	40	–	40	–	–
6e Cuirassiers	–	–	–	40	–
4e Dragons	–	–	–	–	60
2e Div Légère Mécanique					
13e Dragons	40	–	40	–	–
29e Dragons	40	–	40	–	–
8e Cuirassiers	–	–	–	40	–
1er Dragons	–	–	–	–	60
3e Div Légère Mécanique					
1er Cuirassiers	–	40	40	–	–
2e Cuirassiers	–	40	40	–	–
12e Cuirassiers	–	–	–	40	–
11e Dragons	60	–	–	–	–
1re Div Légère de Cavalerie					
1er RAM	12	–	–	12	–
5e Dragons	–	–	–	–	20
2e Div Légère de Cavalerie					
2e RAM	12	–	–	12	–
3e Dragons	–	–	–	–	20
3e Div Légère de Cavalerie					
3e RAM	12	–	3	12	–
2e Dragons	–	–	–	–	–
4e Div Légère de Cavalerie					
4e RAM	12	–	–	12	–
14e Dragons	–	–	–	–	20
5e Div Légère de Cavalerie					
5e RAM	–	12	–	12	–
15e Dragons	–	–	–	–	20
1er GRDI (1DIM)	–	–	–	12	20
2e GRDI (2 DIM)	20	–	–	12	–
3e GRDI (3 DIM)	–	–	–	12	20
4e GRDI (9 DIM)	–	–	–	12	20
5e GRDI (12 DIM)	20	–	–	12	–
6e GRDI (15 DIM)	–	–	–	12	20
7e GRDI (25 DIM)	–	–	–	12	20

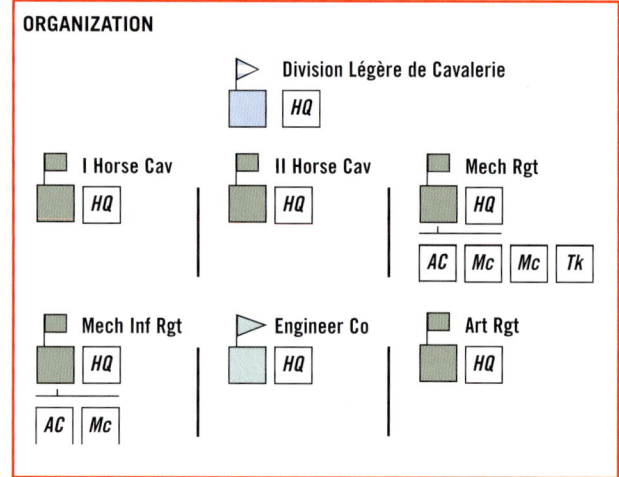

Division Légère de Cavalerie	Strength
Regiments of Horsed Cavalry x 2	
Mechanized Regiment	
Motorcycle Squadrons x 2	
P-178s	12
H-35/H-39s	12
Mechanized Infantry Regiment	
AMR-33/AMR-35s	12
Motorcycle Squadron x 1	
Artillery Regiment	
75mm (2.9in) Field Guns	12
105mm (4.1in) Howitzers	12
Engineer Company x 1	

FRANCE & THE LOW COUNTRIES

was an even more formidable opponent than the Somua, due to its thicker armour and heavier armament. It was only the usual combination of failings in the Char B-equipped units (poor tactics, lack of radios and the one-man turrets) that allowed the Panzers to win through.

Most of the major French armoured formations were now effectively crippled and the German commanders (especially Guderian) exploited the freedom of movement that this provided by driving hard for the Channel coast, covering up to 90km (56 miles) per day until they reached the sea near Abbeville on 20 May. The first phase of the German plan had been completed and it seemed that the Allied units trapped in Belgium were finished (Holland had surrendered on 15 May).

Taking stock

The Panzers had won a spectacular victory, but the practical difficulties of dealing with the heavily armoured French tanks had come as a shock. The German 37mm (1.5in) tank and anti-tank guns had proved to be almost useless against not only the massive Char B but also the Somua and they were only marginally effective against the R-35/H-35 light tanks. In contrast, the 47mm (1.9in) guns of the Char B and Somua were easily capable of penetrating contemporary Panzers, none of which had armour more than 30mm (1.18in) thick. (The first production models of the Stug III assault gun with 50mm/2in frontal armour were just entering service, but only four batteries fielding a total of 24 vehicles took part in the French campaign.) These experiences accelerated existing plans to up-armour serving and future Panzers and to replace the 37mm (1.5in) tank and anti-tank guns with 50mm (2in) weapons.

The French were equally shocked at the speed of the German advance and had the appallingly difficult task of restructuring their forces in the midst of a fiercely fought campaign. The best-equipped French

▲ **Char D2**
The Char D2 was intended as the main battle tank for the DCRs, but only 100 were produced before it was replaced by the more heavily armed and better-armoured Char B.

FRANCE & THE LOW COUNTRIES

◀ **Char de Cavalerie 38H (H-39)**

3e DLM / 1er Cuirassiers

H-39 units were given priority in the up-gunning programme in 1939–40 and a high proportion of their vehicles had been refitted with the L/35 37mm (1.5in) SA-38 gun by the time of the German invasion.

Specifications

Crew: 2
Weight: 12.1 tonnes (11.9 tons)
Length: 4.23m (13ft 10in)
Width: 1.96m (6ft 5in)
Height: 2.16m (7ft 1in)
Engine: 89.5kW (120hp) Hotchkiss 1938 6-cylinder petrol
Speed: 36.5km/h (22.7mph)
Range: 150km (93 miles)
Armament: 1 x 37mm (1.5in) SA-38 gun, plus 1 x coaxial 7.5mm (0.295in) MG

Specifications

Crew: 4
Weight: 32.5 tonnes (32 tons)
Length: 6.52m (21ft 5in)
Width: 2.5m (8ft 2in)
Height: 2.79m (9ft 2in)
Engine: 223.7kW (300hp) Renault 6-cylinder petrol
Speed: 28km/h (17mph)
Range: 180km (112 miles)
Armament: 1 x 75mm (2.9in) ABS SA-35 howitzer in the hull, 1 x 47mm (1.9in) SA-35 gun in turret, plus 2 x 7.5mm (0.295in) MGs (1 coaxial, 1 fixed forward-firing in hull)
Radio: ER51 or ER53

▲ **Char B1 bis**

2e DCR / 8e Bataillon de Chars de Combat

Although the B1 bis was handicapped by its one-man turret, the combination of the turreted 47mm (1.9in) gun and hull-mounted 75mm (2.9in) howitzer was formidable by the standards of 1940.

▲ **Unic P107 Halftrack**

1re DLM / 74e Regiment d'Artillerie

This vehicle was designed by Citroen in the early 1930s as an artillery tractor, but most production was undertaken by Unic. Over 2000 of these vehicles were in service in 1940.

Specifications

Crew: 1 plus up to 6 passengers
Weight: 2.35 tonnes (2.31 tons)
Length: 4.85m (15ft 11in)
Width: 1.8m (5ft 11in)
Height: 1.95m (6ft 4in)
Engine: 41kW (55hp) 4-cylinder petrol
Speed: 45km/h (28mph)
Range: 400km (248.5 miles)

armies, including the most effective armoured formations, had been sent north and were lost in the resulting encirclement.

A front from Sedan to the Channel now had to be defended with greatly reduced forces – 64 French divisions were available, together with a single British division, the 51st Highland Division. They had to cover a front of 600km (375 miles), which theoretically required 60 divisions, leaving a totally inadequate reserve. If the Germans broke through at any point along this line, it would be almost impossible to stop them.

▲ Char 2C (FCM-2C)
51e Bataillon de Chars de Combat

The Char 2C was designed in 1917–18 as a 'breakthrough tank'. An order for 300 vehicles was placed in February 1918, but only 10 were finally completed between 1919 and 1921. It seems probable that by 1940 only eight tanks were still operational and all but one of these were destroyed by their crews on their rail transport wagons to avoid capture.

Specifications

Crew: 12
Weight: 69 tonnes (67.9 tons)
Length: 10.26m (33ft 8in)
Width: 2.95m (9ft 8in)
Height: 4m (13ft 1in)
Engine: 2 x 186.25kW (250hp) Daimler or Maybach 6-cylinder petrol
Speed: 12km/h (7.5mph)
Range: 100km (62 miles)
Armament: 1 x 75mm (2.9in) gun, plus 4 x 8mm (0.315in) Hotchkiss MGs (1 in rear sub-turret, 1 ball-mounted in hull front, 1 ball-mounted each side of hull)
Radio: ER53

▶ Char Léger FCM-36
7e Bataillon de Chars de Combat

The FCM-36 was designed by the Société des Forges et Chantiers de la Mediterranée for the infantry support role. The 100 vehicles produced were only issued to the 4e and 7e BCC. The design incorporated sloped armour and an octagonal FCM turret but was poorly armed with an L/21 37mm (1.5in) SA-18 gun and a coaxial machine gun. (A very small number may have been rearmed with the L/33 37mm/1.5in SA-38 gun.)

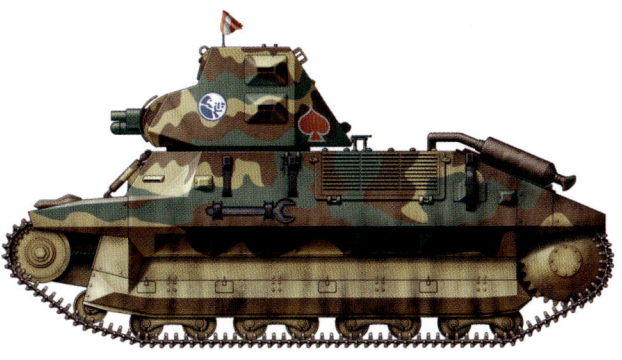

Specifications

Crew: 2
Weight: 12.35 tonnes (12.15 tons)
Length: 4.22m (13ft 10in)
Width: 1.95m (6ft 4.75in)
Height: 2.15m (7ft 0.61in)
Engine: 67.9kW (91hp) Berliet 4-cylinder diesel
Speed: 24km/h (15mph)
Range: 225km (140 miles)
Armament: 1 x 37mm (1.5in) SA-18 gun, plus 1 x 7.5mm (0.295in) coaxial MG

 The battalion insignia of *7e Bataillion de Chars de Combat* shows a stylized depiction of a tank gunner. The red spade on the rear turret indicates that this tank is from the 1st section.

FRANCE & THE LOW COUNTRIES

Counterattacks
May 1940

The German forces that had reached the Channel coast on 20 May sought to establish bridgeheads across the River Somme and to tighten the ring around the Allied armies trapped against the Channel.

However, they were themselves dependent on a long, narrow corridor stretching back to the River Meuse. Until the infantry with their horse-drawn artillery and supply columns could catch up with the Panzers, there was still a chance for the Allies to turn the tables. Forces holding the line of the Somme tried to eliminate the German bridgeheads on the south bank. These forces included the newly arrived but under-strength British 1st Armoured Division. Its initial operations west of Amiens on the 24th were

▲ **Char B1 bis**
3e DCR / 41e Bataillon de Chars de Combat
When boldly handled, the B1 bis was truly formidable. On 16 May 1940, at Stonne, a single vehicle ('Eure' commanded by Lieutenant Bilotte) attacked a column of tanks of 8th Panzer Regiment, destroying two Panzer IVs, 11 Panzer IIIs and two 37mm (1.5in) anti-tank guns. The B1 bis took an estimated 140 hits from return fire, none of which penetrated the armour.

Specifications
Crew: 4
Weight: 32.5 tonnes (32 tons)
Length: 6.52m (21ft 5in)
Width: 2.5m (8ft 2in)
Height: 2.79m (9ft 2in)
Engine: 223.7kW (300hp) Renault 6-cylinder petrol
Speed: 28km/h (17mph)
Range: 180km (112 miles)
Armament: 1 x 75mm (2.9in) ABS SA-35 howitzer in the hull, 1 x 47mm (1.9in) SA-35 gun in turret, plus 2 x 7.5mm (0.295in) MGs (1 coaxial, 1 fixed forward-firing in hull)
Radio: ER51 or ER53

 Some French tank battalions and regiments used colourful symbols – *4e Régiment de Cuirassiers*, for example, adopted a red and white Joan of Arc badge.

▶ **Char de Cavalerie 35H (H-35)**
1re DLM / 4e Cuirassiers
This H-35 was one of the relatively small number of radio-equipped French tanks in 1940. Widespread reliance on verbal briefings and flag signals played an important part in slowing the tempo of French armoured operations.

Specifications
Crew: 2
Weight: 10.6 tonnes (10.43 tons)
Length: 4.22m (13ft 10in)
Width: 1.96m (6ft 5in)
Height: 2.62m (8ft 7in)
Engine: 55.91kW (75hp) Hotchkiss 1935, 6-cylinder petrol
Speed: 27km/h (17mph)
Range: 150km (93 miles)
Armament: 1 x 37mm (1.5in) SA-18 gun, plus 1 x coaxial 7.5mm (0.295in) MG
Radio: ER51

FRANCE & THE LOW COUNTRIES

partially successful, but the ground gained could not be held against increasing German resistance. Further attacks, supported by the re-formed 2e and 5e DLCs, were mounted on 27 May in which 1st Armoured Division lost 110 of its 257 tanks. French armoured units, including the rebuilt 2e DCR, continued the attacks until 4 June, by which time they had suffered such heavy losses that they were no longer effective fighting units. These losses sealed the fate of the Weygand Line, the defences along the Rivers Somme and Aisne that were intended to buy time to rebuild the battered French forces. In themselves, these defences were highly effective, comprising deep belts of fortified villages and other mutually supporting strongpoints. Their crucial weakness lay in the lack of adequate mobile

Specifications

Crew: 3
Weight: 19.5 tonnes (19.2 tons)
Length: 5.38m (17ft 7.8in)
Width: 2.12m (6ft 11.5in)
Height: 2.62m (8ft 7in)
Engine: 141.7kw (190hp) V8 petrol
Speed: 41km/h (25mph)
Range: 257km (160 miles)
Armament: 1 x 47mm (1.9in) SA-35 gun, plus 1 x coaxial 7.5mm (0.295in) MG

▲ Char de Cavalerie 35S (Somua)
1re DLM / 18e Dragons

The use of national roundels on French tanks became widespread at the time of the German invasion after several 'friendly fire' incidents, such as that on 19 May 1940, when Char Bs of the newly formed 4e DCR took several casualties from the fire of Somuas of 3e Cuirassiers.

▲ Char B1 bis
4e DCR / 47e Bataillon de Chars de Combat

The B1 bis rapidly acquired a fearsome reputation amongst German troops from its ability to shrug off repeated hits from 37mm (1.5in) tank and anti-tank guns. The tanks were often referred to as: *Stahlkolosse* ('iron colossus'), *Stahlriesen* ('iron giants'), *Stahlfestungen* ('iron fortresses'), *stählerne Kasten* ('iron boxes'), *Riesentiere* ('giant beasts'), *Ungeheuer*, *Ungetüme* or *Untiere* ('monsters').

Specifications

Crew: 4
Weight: 32.5 tonnes (32 tons)
Length: 6.52m (21ft 5in)
Width: 2.5m (8ft 2in)
Height: 2.79m (9ft 2in)
Engine: 223.7kW (300hp) Renault 6-cylinder petrol
Speed: 28km/h (17mph)
Range: 180km (112 miles)
Armament: 1 x 75mm (2.9in) ABS SA-35 howitzer in the hull, 1 x 47mm (1.9in) SA-35 gun in turret, plus 2 x 7.5mm (0.295in) MGs (1 coaxial, 1 fixed forward-firing in hull)
Radio: ER51 or ER53

FRANCE & THE LOW COUNTRIES

forces to seal off any penetration of the line. General Maxime Weygand concentrated his best remaining units in the coastal sector, where he expected the Germans to strike for the ports to cut off reinforcements and supplies from Britain, and on the plain of Champagne east of Reims, which was ideal terrain for armoured operations. He had correctly assessed German intentions, but the entire Somme portion of the new line was weak from the outset because of German bridgeheads established during their advance to the Channel.

On 5 June, Army Group B launched what was considered a secondary effort north-west of Paris. Although French resistance was fierce, German reinforcements were instrumental in achieving a decisive breakthrough on the 8th. Success was largely

▲ **Char B1 bis**
4e DCR / 46e Bataillon de Chars de Combat
Primarily designed as a close support weapon for firing high explosive (HE) against anti-tank guns and field fortifications, the B1 bis' L/17.1 75mm (2.9in) SA-35 howitzer also had a useful armour-piercing high explosive (APHE) shell that could penetrate 40mm (1.57in) of armour at 400m (438 yards).

Specifications
Crew: 4
Weight: 32.5 tonnes (32 tons)
Length: 6.52m (21ft 5in)
Width: 2.5m (8ft 2in)
Height: 2.79m (9ft 2in)
Engine: 223.7kW (300hp) Renault
   6-cylinder petrol
Speed: 28km/h (17mph)
Range: 180km (112 miles)
Armament: 1 x 75mm (2.9in) ABS SA-35 howitzer in the hull, 1 x 47mm (1.85in) SA-35 gun in turret, plus 2 x 7.5mm (0.295in) MGs (1 coaxial, 1 fixed forward-firing in hull)
Radio: ER51 or ER53

▲ **Char Léger 17R (Renault FT-17)**
Armée des Alpes / Bataillon Colonial
Over 3700 FT-17s were built in French factories between 1917 and 1921, almost 1300 of which were still in service in 1940. Eight tank battalions each fielded 63 FT-17s and the type also equipped some independent tank companies.

Specifications
Crew: 2
Weight: 6.6 tonnes (6.5 tons)
Length: 4.09m (13ft 5in)
Width: 1.7m (5ft 7in)
Height: 2.13m (7ft)
Engine: 26.1kw (35hp) Renault
   4-cylinder petrol
Speed: 7.7km/h (4.8mph)
Range: 35km (21.7 miles)
Armament: 1 x 37mm (1.5in) SA-18 gun or 1 x 7.5mm (0.295in) MG

FRANCE & THE LOW COUNTRIES

due to the lack of fully combat-worthy major French armoured formations. The 2e, 3e and 4e DCRs could field only 150 tanks between them, whilst the newly raised 7e DLM had a further 174 AFVs.

As the French forces north-west of Paris retreated, they exposed the left flank of the armies defending the line of the Aisne. It was in this sector that General Gerd von Rundstedt's Army Group A launched the German main effort on 9 June. It managed only limited advances during the first three days in the face of repeated French counterattacks, but the constant German pressure and an increasingly exposed left flank forced the defenders to pull back to the Marne on the 11th. The next day Guderian's four Panzer divisions penetrated this flimsy line and rapidly exploited the breakthrough. Although the remnants of the French armoured formations fought a series of fierce small-scale actions until 16 June, they could do no more than delay the German advance, which only ended with the French surrender on the 22nd.

Specifications
Crew: 3
Weight: 6.91 tonnes (6.8 tons)
Length: 4.83m (16ft 1in)
Width: 1.73m (5ft 9.5in)
Height: 2.6m (8ft 8in)
Engine: 44.7kW (60hp) Panhard 17 petrol
Speed: 50km/h (31mph)
Range: 251km (156 miles)
Armament: 1 x 37mm (1.5in) SA-18 gun, plus 1 x coaxial 7.5mm (0.295in) MG

▲ **AMC Schneider P16**
1er Groupe de Reconnaissance de Division d'Infanterie
Roughly 100 P16s were produced from 1928 to 1930. They originally served with cavalry divisions in *Escadrons d'Automitrailleuses de Combat* (EAMCs), but were subsequently transferred to five of the GRDIs, each of which had 12 vehicles. The remaining serviceable P16s were sent to North Africa to operate in support of the *Chasseurs d'Afrique*.

Specifications
Crew: 1
Weight: 2.85 tonnes (2.8 tons)
Length: 4.64m (15ft 2.7in)
Width: 1.85m (6ft 1in)
Height: 2.15m (7ft 1in)
Engine: 40.98kW (55hp) 4-cylinder petrol
Speed: 51km/h (32mph)
Range: n/k

▲ **Laffly S15T**
German occupation forces, France 1943
The Laffly S15T was a French artillery tractor designed to tow light weapons such as the 75mm (2.9in) field gun and the 105mm (4.1in) howitzer. Captured examples were widely used by German occupation forces.

FRANCE & THE LOW COUNTRIES

▲ Lorraine 38L (VBCP)
1re DCR / Bataillon de Chasseurs Portés

The Lorraine 38L – also known as a VBCP (*Voiture Blindée de Chasseurs Portés*, or 'Armoured Mounted Infantry Vehicle') – was the world's first fully tracked armoured personnel carrier to go into action. Orders were placed for 440 vehicles, but only 150 were delivered by May 1940 and most of these were issued to the DCRs' mechanized chasseur (infantry) battalions on a scale of 40 per battalion. The Lorraine 38L could carry six men in the tractor, with a further six in the armoured trailer and most vehicles were fitted with one or two AA mountings.

Specifications
Crew: 2 plus 8 passengers
Weight: 6.2 tonnes (6.1 tons)
Length: 4.22m (13ft 10in)
Width: 1.57m (5ft 3in)
Height: 2.13m (7ft) (estimated)
Engine: 52.15kW (70hp) Delahaye Type 135 6-cylinder petrol
Speed: 35km/h (22mph)
Range: 137km (85 miles)

▲ Citroen-Kegresse P19 (CK P19)
1re DLC / 5e Dragons

The P19 entered production in 1932 and a total of 547 remained in service in 1940. Most vehicles were used as personnel carriers in French cavalry formations.

Specifications
Crew: 5–7
Weight: 4.05 tonnes (3.9 tons)
Length: 4.85m (15ft 11in)
Width: 1.8m (5ft 11in)
Height: 1.95m (6ft 5in)
Engine: 41kW (55hp) 4-cylinder petrol
Speed: 45km/h (28mph)
Range: 400km (248.5miles)

▶ Panhard 178 Armoured Car
German occupation forces, France 1942

In June 1941, a total of 190 Panhards were issued to German units. Forty-three were fitted with flanged steel wheels for use on railways. In 1943, some were fitted with a new open-topped turret mounting a 50mm (2in) L/42 or L/60.

Specifications
Crew: 3
Weight: 8.63 tones (8.5 tons)
Length: 4.79m (15ft 8in)
Width: 2.01m (6ft 7in)
Height: 2.31m (7ft 7in)
Engine: 78kW (105hp) Panhard SK 4-cylinder petrol
Speed: 72km/h (45mph)
Range: 300km (186 miles)
Armament: 1 x 25mm (1in) gun, plus 1 x coaxial 7.5mm (0.295in) MG
Radio: ER 26 or ER29

FRANCE & THE LOW COUNTRIES

BEF in France
May 1940

In 1918 the British Army had built up great expertise in the battlefield handling of AFVs and was now rapidly developing a wide range of special-purpose vehicles.

SUPPLY TANKS AND GUN carriers that could transport medium artillery had been in service as early as 1917 whilst prototypes of APCs and engineer tanks were being trialled. At Amiens in August 1918, a total of almost 450 British AFVs inflicted 26,000 casualties on the Germans and captured 400 guns, scoring an even greater success than that achieved by the French at Soissons the previous month. In recognition of the value of AFVs, the Tank Corps had grown to a total of strength of 25 battalions by the Armistice of November 1918.

Post-war austerity

In the first years of peace a rapid run-down in tank numbers meant that by 1921 only five battalions and 12 armoured car companies remained. Although there was often great pressure to bring AFVs under infantry control, the Royal Tank Corps survived and during the early 1920s received its first major new tank type, the Vickers Medium.

Although it was very thinly armoured, by the standards of the time the Vickers Medium was a fast and reliable tank that played a vital role in testing the theories of armoured warfare during the 1920s and 1930s. The most important of these tests came with the formation of the Experimental Mechanised Force in 1927.

This was the ancestor of all armoured divisions and in a series of exercises it showed the potential battle-winning capabilities of such formations. Foreign military observers were deeply impressed and

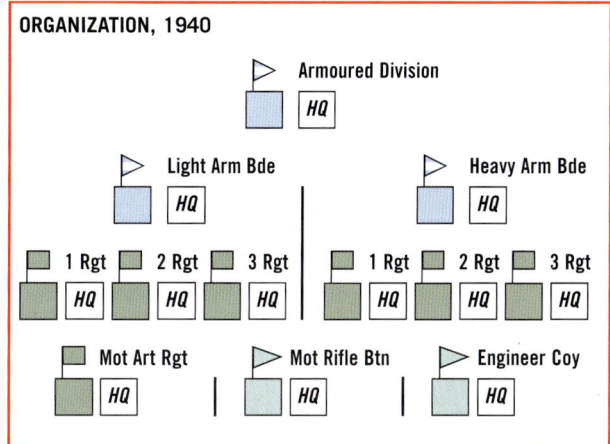

ORGANIZATION, 1940

BEF Armoured Units, 10 May 1940	Mark VI	Matilda I	Matilda II	Cruiser Tank	Carrier	Daimler	Morris	Guy
4th/7th Dragoon Guards	28	–	–	–	44	–	–	–
5th Dragoon Guards	28	–	–	–	44	–	–	–
13th/18th Hussars	28	–	–	–	44	–	–	–
15th/19th Hussars	28	–	–	–	44	–	–	–
1st Lothians	28	–	–	–	44	–	–	–
1st Fife and Forfar	28	–	–	–	44	–	–	–
East Riding Yeomanry	28	–	–	–	44	–	–	–
12th Lancers	–	–	–	–	–	–	38	–
4th Northumberland Fusiliers (infantry)	–	–	–	–	–	12	–	–
1st Army Tank Bde								
4 RTR (Royal Tank Regiment)	5	50	–	–	8	–	–	–
7 RTR	7	27	23	–	8	–	–	–
No. 3 Air Mission Phantom	–	–	–	–	–	–	–	6

FRANCE & THE LOW COUNTRIES

▲ **Training**
Mark VIB light tanks and Scout carriers of the BEF participate in manoeuvres during the winter of 1939/40.

their reports influenced the development of German, Soviet and United States armoured units.

By the late 1930s rearmament programmes were slowly beginning to produce desperately needed new tanks and enough were available for the first two Mobile Divisions to form in 1938. In 1940 these formations were retitled, becoming 1st and 7th Armoured Divisions.

Separate role

The armoured divisions (one based in the United Kingdom and the other, the 7th, later the 'Desert Rats', in Egypt) were not designed for the infantry support role, which was assigned to specialist army tank brigades, each of which was supposed to have three regiments equipped with heavily armoured infantry tanks. Infantry divisions had some light AFVs of their own in the form of Bren, Scout and (later) Universal Carriers (a total of 96 per division). A mechanized cavalry regiment with 28 light tanks and 44 carriers was also attached to each infantry division to provide close reconnaissance.

When the German offensive opened on 10 May 1940, the BEF's principal armoured units were the 1st Army Tank Brigade (77 Matilda Is, 23 Matilda IIs and 12 Mark VI light tanks, plus 16 carriers) and eight mechanized cavalry regiments, seven with the mix of light tanks and carriers given above and an eighth with 38 Morris CS9/LAC armoured cars.

INSIGNIA

British vehicles were khaki green, over-painted with an irregular disruptive pattern of dark green.

INSIGNIA

The most conspicuous tactical markings were a solid white recognition square (see top left) about 30cm x 30cm (12in x 12in) carried on the hull front, rear and sides, and company/squadron outlined symbols, which were applied to each side of the turret (or hull side for non-turreted vehicles).

These latter symbols were:

Diamond	HQ
Triangle	A Company or Squadron (see above)
Square	B Company or Squadron
Circle	C Company or Squadron

Although these symbols were supposed to be colour-coded, in practice a wide variety of colours were used, including red, white, yellow, blue and purple.

FRANCE & THE LOW COUNTRIES

▲ Bedford OYD GS 3-Ton Truck
1st Armoured Division / 2nd Armoured Brigade / 10th Royal Hussars / HQ Squadron / F Echelon

Over 72,000 OY series '3-Tonners' were produced between 1939 and 1945. Officially, each armoured regiment fielded 22 such vehicles.

Specifications
Crew: 1
Weight: 6.57 tonnes (6.46 tons)
Length: 6.22m (20ft 4.9in)
Width: 2.18m (7ft 2in)
Height: 3.09m (10ft 1.6in)
Engine: 53.64kW (72hp) Bedford WD 6-cylinder petrol
Speed: 72.5km/h (45mph) (estimated)
Range: 450km (280 miles)

▶ Light Tank Mark VIC
1st Armoured Division / 2nd Armoured Brigade / 10th Royal Hussars

The Mark VIC was fitted with a new turret, mounting a 15mm (0.59in) Besa heavy machine gun and a coaxial 7.92mm (0.31in) Besa machine gun.

Specifications
Crew: 3
Weight: 5.08 tonnes (5 ton)
Length: 4.01m (13ft 2in)
Width: 2.08m (6ft 10in)
Height: 2.13m (6ft 11in)
Engine: 65.6kW (88hp) Meadows 6-cylinder petrol
Speed: 56km/h (35mph)
Range: 200km (124 miles)
Armament: 1 x 15mm (0.59in) Besa HMG, plus 1 x coaxial 7.92mm (0.31in) Besa MG
Radio: Wireless Set No. 9

▶ Light Tank Mark VIB
2nd Infantry Division / 4th/7th Royal Dragoon Guards

Over 200 of these tanks were in service with the BEF in May 1940. Whilst they were effective as reconnaissance vehicles, their thin armour and feeble armament caused heavy losses when they were forced to act in the role of 'substitute cruiser or infantry tank'.

Specifications
Crew: 3
Weight: 5.08 tonnes (5 ton)
Length: 4.01m (13ft 2in)
Width: 2.08m (6ft 10in)
Height: 2.26m (7ft 5in)
Engine: 65.6kW (88hp) Meadows 6-cylinder petrol
Speed: 56km/h (34.78mph)
Range: 200km (124.2 miles)
Armament: 1 x 12.7mm (0.5in) Vickers HMG, plus 1 x coaxial 7.62mm (0.3in) Vickers MG
Radio: Wireless Set No. 9

FRANCE & THE LOW COUNTRIES

▲ Cruiser Tank Mark IV, A13 Mark II
1st Armoured Division / 2nd Armoured Brigade / 2nd Dragoon Guards (Queen's Bays)

Despite being used for tasks better suited to heavily armoured infantry tanks, the cruiser tanks of 1st Armoured Division scored some notable tactical successes before being overwhelmed by superior German weaponry and tactics.

Specifications
Crew: 4
Weight: 15.04 tonnes (14.8 tons)
Length: 6.02m (19ft 9in)
Width: 2.59m (8ft 6in)
Height: 2.59 (8ft 6in)
Engine: 253.64kW (340hp) Nuffield Liberty V12 petrol
Speed: 48km/h (30mph)
Range: 145km (90 miles)
Armament: 1 x 40mm (1.57in) 2pdr Ordnance Quick-Firing (OQF) gun, plus 1 x coaxial 7.92mm (0.31in) Besa MG
Radio: Wireless Set No. 9

◀ Morris CS9 Armoured Car
GHQ BEF / 12th Royal Lancers

The Morris CS9 armoured car was based on the chassis of the Morris 4x2 15cwt truck. A total of 99 were built in 1938, of which 38 equipped the 12th Royal Lancers, the only armoured car unit in the BEF.

Specifications
Crew: 4
Weight: 4.26 tonnes (4.2 tons)
Length: 4.78m (15ft 8in)
Width: 2.06m (6ft 9in)
Height: 2.21m (7ft 3in)
Engine: 52kW (70hp) Morris Commercial 4-cylinder petrol
Speed: 72km/h (45mph)
Range: 386km (240 miles)
Armament: 1 x 14mm (0.55in) Boys anti-tank rifle, plus 1 x 7.7mm (0.303in) Bren MG
Radio: Wireless Set No. 9

▲ Morris CS8 GS 15cwt Truck
1st Armoured Division / 2nd Armoured Brigade / 2nd Dragoon Guards (Queen's Bays) / HQ Squadron / F Echelon

The CS8 entered service in 1934 and over 21,000 vehicles were built before production ceased in 1941. A total of 12 GS 15cwt trucks were held by each armoured regiment. Many were captured in May/June 1940 and were taken into service with the *Wehrmacht*.

Specifications
Crew: 1
Weight: 0.99 tonnes (0.9 tons)
Length: 4.80m (15ft 9in)
Width: 1.80m (5ft 11in)
Height: 1.85m (6ft 1in)
Engine: 3.5 litre (213ci) 4-cylinder petrol
Speed: 75km/h (47mph)
Range: 225km (136 miles)
Radio: None

FRANCE & THE LOW COUNTRIES

▶ Light Tank Mark VIB
1st Armoured Reconnaissance Brigade / East Riding Yeomanry

Only a handful of the 300-plus British light tanks deployed to France were successfully evacuated, contributing to Britain's desperate shortage of AFVs in the summer of 1940.

Specifications
Crew: 3
Weight: 5.08 tonnes (5 tons)
Length: 4.01m (13ft 2in)
Width: 2.08m (6ft 10in)
Height: 2.26m (7ft 5in)
Engine: 65.6kW (88hp) Meadows 6-cylinder petrol
Speed: 56km/h (34.78mph)
Range: 200km (124 miles)
Armament: 1 x 12.7mm (0.5in) Vickers HMG, plus 1 x coaxial 7.62mm (0.303in) Vickers MG
Radio: Wireless Set No. 9

▶ Bren Gun Carrier
2nd Infantry Division / 4th Brigade / 1st Battalion The Royal Scots

Each of the BEF's infantry divisions had a total of 96 Bren and Scout Carriers. These were scheduled to be replaced by the more versatile Universal Carrier, but few, if any, of these new vehicles reached the BEF before the Dunkirk evacuation.

Specifications
Crew: 3
Weight: 3.81 tonnes (3.75 tons)
Length: 3.65m (11ft 11in)
Width: 2.05m (6ft 9in)
Height: 1.45m (4ft 9in)
Engine: 52.2kW (70hp) Ford 8-cylinder petrol
Speed: 48km/h (30mph)
Range: 210km (130 miles)
Armament: 1 x 14mm (0.55in) Boys anti-tank rifle, plus 1 x 7.7mm (0.303in) Bren MG or 2 x 7.7mm (0.303in) Bren MGs

Counterattack at Arras
21 May 1940

Both sides of the 'Panzer corridor' from the Meuse to the Channel were potentially vulnerable to counterattacks. Repeated attempts were made by the Allies to break through from the south, but it was the Arras counterattack launched by the forces trapped against the Channel coast that was to have the greatest impact on the outcome of the campaign.

IN MANY RESPECTS, the Arras counterattack was improvised rather than planned. It was launched on 21 May by 58 Matilda Is and 16 Matilda IIs of the BEF's 1st Army Tank Brigade, supported by the remaining 60 Somuas of 3e DLM. Their advance hit the SS Motorized Infantry Regiment *Totenkopf* and 7th Panzer Division's rifle battalions, who found to their horror that the Matildas were immune to the fire of their 37mm (1.5in) anti-tank guns. As the tanks pushed on into the German rear, General Erwin Rommel frantically formed a defence line using 7th Panzer Division's 88mm (3.5in) Flak guns and its medium artillery. These weapons were decisive as the British armour had outrun its supporting infantry and artillery. Eight Matildas were knocked out and the rest were forced to withdraw, but when the Panzer 38(t)s of 25th Panzer Regiment attempted to cut off the retreat, they ran into a screen of 2pdr anti-tank guns backed up by Somuas and were beaten off. A total of about 40 Matildas were lost in

FRANCE & THE LOW COUNTRIES

exchange for six Panzer 38(t)s, three Panzer IVs and many lighter AFVs and anti-tank guns.

The importance of this action went far beyond the losses on either side – Rommel was convinced that further attacks were imminent and halted his advance for 24 hours, giving the British time to reinforce the vital Channel coast, especially the ports of Boulogne and Calais, which held out against XIX Panzer Corps until 25 and 27 May respectively. Coupled with reports of continuing small-scale French counter-attacks, the Arras fighting raised fears amongst many of the more senior German generals that the Panzer divisions might be cut off from the infantry divisions who were making epic forced marches to catch up with them. Understandably, most were relieved when, on 24 May, Hitler ordered the Panzer divisions out of the line to allow time for them to refit before the next stage of the campaign. Although Guderian protested strongly against this ruling, it was not rescinded until two days later, by which point the Dunkirk evacuation (Operation *Dynamo*) was about to begin. The Allied defences were able to hold the renewed German attacks until 4 June, by which time over 330,000 troops (one third of them French) had been rescued, although virtually all their AFVs, artillery and heavy equipment had to be abandoned.

▲ **Matilda I Infantry Tank**
1st Army Tank Brigade / 4 RTR
A total of 77 Matilda Is were issued to the 1st Army Tank Brigade by May 1940. The armour of the 58 vehicles that took part in the Arras counterattack proved to be virtually immune to the German 37mm (1.5in) anti-tank gun, although the vehicles' exposed tracks were vulnerable.

Specifications
Crew: 2
Weight: 11.16 tonnes (11 tons)
Length: 4.85m (15ft 11in)
Width: 2.29m (7ft 6in)
Height: 1.85m (6ft 1in)
Engine: 52.2kW (70hp) Ford V8 petrol
Speed: 13km/h (8mph)
Range: 129km (80 miles)
Armament: 1 x 12.7mm (0.5in) Vickers HMG or 1 x 7.7mm (0.303in) Vickers MG
Radio: Wireless Set No. 9

▲ **Matilda II Infantry Tank**
1st Army Tank Brigade / 7 RTR
Only 16 of 7 RTR's 23 Matilda IIs were available for the Arras counterattack, but their apparent invulnerability and aggressive handling had an immense psychological impact on the German chain of command.

Specifications
Crew: 4
Weight: 26.92 tonnes (26.5 tons)
Length: 5.61m (18ft 5in)
Width: 2.59m (8ft 6in)
Height: 2.52m (8ft 3in)
Engine: 2 x 64.8kW (87hp) AEC 6-cylinder diesel
Speed: 13km/h (8mph)
Range: 258km (160 miles)
Armament: 1 x 40mm (1.57in) 2pdr Ordnance Quick-Firing (OQF) gun, plus 1 x coaxial 7.92mm (0.31in) Besa MG
Radio: Wireless Set No. 11

FRANCE & THE LOW COUNTRIES

▼ Counterattack force at Arras

The Allied tank forces that formed the main striking force of the Arras counterattack were an ad hoc collection of units, with no experience of working together. Battlefield procedures were improvised and unreliable communications hampered the co-ordination of the disparate elements of the force. In the circumstances, their achievements were truly remarkable.

4 RTR at Arras (7 x Matilda II, 29 Matilda I)

7 RTR at Arras (9 x Matilda II, 27 Matilda I)

3e DLM (60 Somuas)

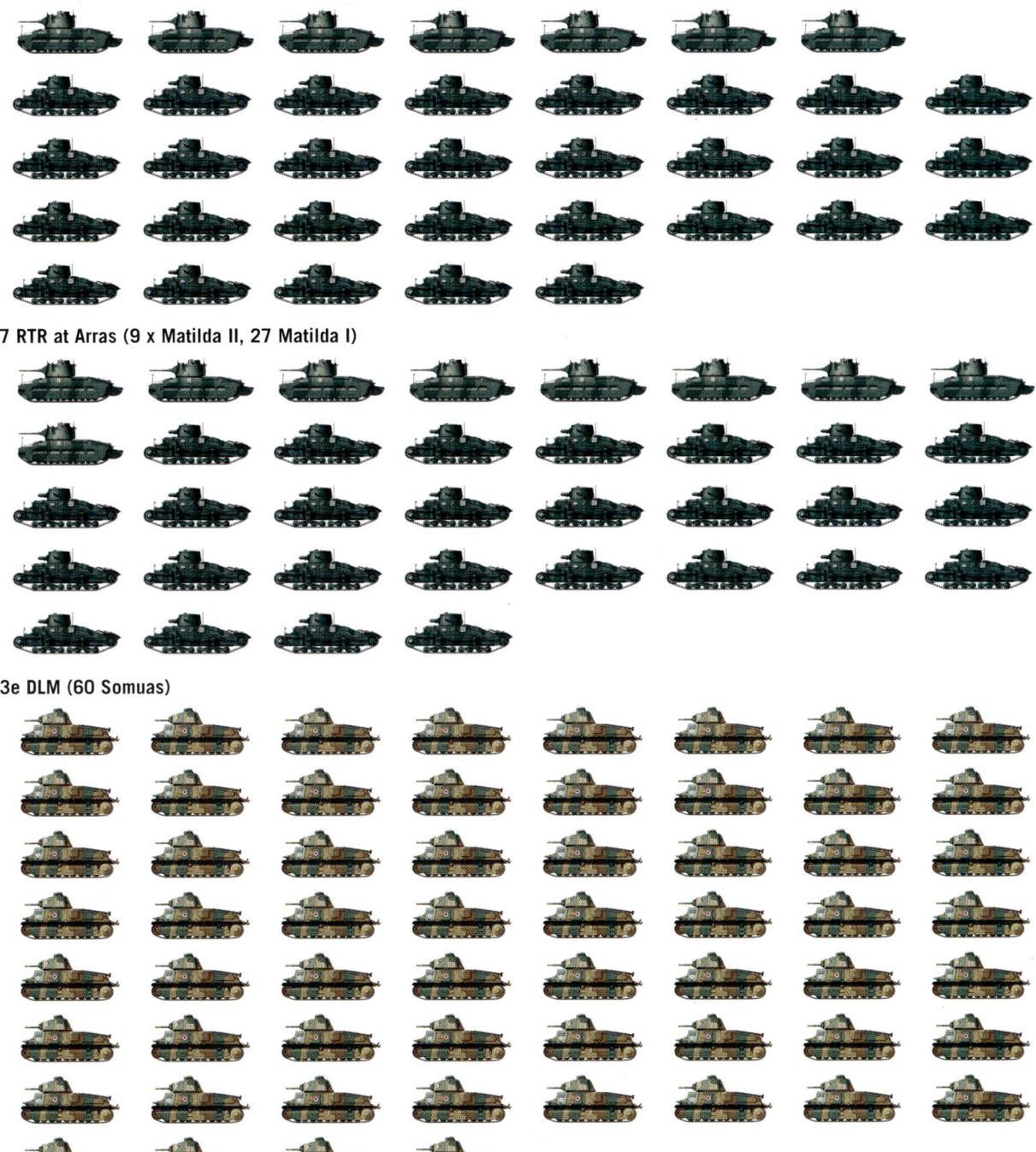

Chapter 3

North Africa: 1940–43

For almost three years, the deserts of North Africa were the setting for some of the most dramatic battles in the history of armoured warfare. They ruined the reputations of a string of British generals, but created two legends. Rommel gained fame as the 'Desert Fox' and acquired the grudging admiration of his opponents, who came to refer to any successful job as 'a Rommel'. Montgomery's victory at the Second Battle of El Alamein made his name as a great commander and set him on the road that would lead to the command of Twenty-first Army Group and a field-marshal's baton.

◀ **Moving up**
A newly delivered Grant en route to the front-line. This vehicle is from an early production batch with the short (L/31) M2 75mm (2.9in) gun. Later Grants and Lees had the longer (L/40) M3 75mm (2.9in) gun.

NORTH AFRICA: 1940-43

Early operations

Italy's conquest of Ethiopia in the mid-1930s raised British fears that Mussolini's ambitions included the seizure of Egypt and the Suez Canal.

A SCRATCH FORCE OF AFVs was assembled at Mersa Matruh during 1936 to act as a deterrent to any hostile move by the large Italian garrison of Cyrenaica, less than 160km (100 miles) away to the west. This British force was slowly strengthened as tensions mounted in Europe during the 1930s until it became the Mobile Division Egypt following the 1938 Munich crisis. The division's first commander was General Percy Hobart, one of the pioneers of armoured warfare, whose fierce training made it the most formidable unit in North Africa. He made high-ranking enemies at HQ in Cairo with his prickly nature, coupled with his insistence that AFVs were the only real battle winners, and he was sacked in September 1939. For all Hobart's faults, the first desert victory was largely due to his old division, which General Richard O'Connor described as 'the best trained division I have ever seen'. Indeed, the division's training was soon to be tested in action, for the French surrender on 22 June 1940 released all 250,000 Italian troops in North Africa for operations against Egypt. They were opposed by the 10,000 British troops of the Western Desert Force – truly daunting odds by any standards.

▶ **Hot pursuit**
Mark VI light tanks follow up the retreating Italian forces, December 1940.

A classic victory – the desert war
June 1940–February 1941

Immediately after Mussolini's declaration of war on 10 June 1940, Morris and Rolls Royce armoured cars of the 11th Hussars began a series of raids into Libya.

THEIR AMBUSHES DISRUPTED convoys and began hitting Italian morale, which took another blow on 28 June when the able and popular Italian commander-in-chief, Air Marshal Italo Balbo, was killed when his aircraft was shot down. He had been demanding vast quantities of equipment before launching the offensive that Mussolini demanded – 100 water tankers, 1000 lorries and more medium tanks and anti-tank guns, all of which were essential.

His replacement, Marshal Rodolfo Graziani, repeated these demands but, despite airy promises, only a fraction of the required equipment was delivered after long delays. Under intense pressure from Mussolini, the much-postponed Italian

British Tank Units, June 1940	Vehicles
2 RTR	A9, A10, A13 cruisers
7 RTR	Matilda II infantry tanks
3rd Hussars	Mark VI light tanks

offensive was finally launched by five divisions on 13 September. The heavily outnumbered British screening forces covering the border fell back, fighting a series of delaying actions, until the Italians halted around Sidi Barrani, about 96km (60 miles) inside Egypt. This halt surprised the British, but Graziani had little choice as his tanks and motor transport were suffering frequent breakdowns whilst

NORTH AFRICA: 1940–43

the infantry, which made up most of his force, were exhausted after three days of marching in the searing heat.

The small-scale armoured actions since the opening of the desert war had shown just how inferior Italian AFVs were to their British counterparts. This was most obvious as far as armament was concerned, for the L3s had no anti-armour weapons, but could be knocked out by the heavy machine guns and anti-tank rifles of the opposing light tanks, carriers and armoured cars. The M11/39 was little better off as its hull-mounted 37mm (1.5in) gun proved to be ineffective against A10 and A13 tanks except at point-blank range.

In contrast, the 2pdr guns of the British cruiser tanks could easily penetrate the M11/39's frontal armour at normal battlefield ranges. Equally important was the scale of issue of radios – almost all British AFVs had them, but very few Italian tanks were so equipped. This meant that Italian armoured operations were slowed by the need to halt for new orders as the tactical situation changed – the only alternative was for junior officers to issue 'follow me' orders – a practice that could cause disaster if they became casualties.

Given these factors, plus the continuing shortages of fuel and artillery ammunition, Graziani's decision to secure his new positions with a line of fortified

▶ **Light Tank Mark VIB**
Western Desert Force / 7th Armoured Division / 7th Armoured Brigade / 1 RTR
In the first months of the desert war, the Mark VI light tanks were found to be markedly superior to their Italian counterparts.

Specifications
Crew: 3
Weight: 5.08 tonnes (5 tons)
Length: 4.01m (13ft 2in)
Width: 2.08m (6ft 10in)
Height: 2.26m (7ft 5in)
Engine: 65.6kW (88hp) Meadows 6-cylinder petrol
Speed: 56km/h (35mph)
Range: 200km (124 miles)
Armament: 1 x 12.7mm (0.5in) Vickers HMG, plus 1 x coaxial 7.62mm (0.3in) Vickers MG
Radio: Wireless Set No. 9

▲ **Cruiser Tank Mark II, A10 Mark IA**
Western Desert Force / 7th Armoured Division / 4th Armoured Brigade / 7th Hussars
The A10's 2pdr gun was capable of destroying any Italian tank at normal battlefield ranges.

Specifications
Crew: 5
Weight: 13.97 tonnes (13.75 tons)
Length: 5.51m (18ft 1in)
Width: 2.54m (8ft 4in)
Height: 2.59m (8ft 6in)
Engine: 111.9kW (150hp) AEC Type 179 6-cylinder petrol
Speed: 26km/h (16mph)
Range: 161km (100 miles)
Armament: 1 x 40mm (1.57in) 2pdr OQF gun, plus 2 x 7.92mm (0.31in) Besa MGs (1 coaxial and 1 ball-mounted in hull front)
Radio: Wireless Set No. 9

NORTH AFRICA: 1940–43

camps is understandable. Greatly exaggerated reports of British reinforcements also played a part in this decision and strengthened his reluctance to attempt any further advance until more of the long-awaited equipment arrived.

Italian caution prompted both General O'Connor, commanding the Western Desert Force at Mersa Matruh, and General Archibald Wavell, commander-in-chief in Cairo, to plan counter-offensives. Since June, their commands had received reinforcements that had greatly increased their fighting power.

Operation *Compass*

Although Graziani still had a vast front-line numerical superiority (80,000 to 30,000 troops), the position was reversed as far as the vital AFVs were concerned, with 275 British tanks opposing 120 greatly inferior Italian vehicles. Preparations for the

▲ **Cruiser Tank Mark II, A10 Mark IA**

Western Desert Force / 7th Armoured Division / 4th Armoured Brigade / 2 RTR

The A10's 30mm (1.18in) frontal armour was proof against most Italian tank guns except at very close range.

Specifications
Crew: 5
Weight: 13.97 tonnes (13.75 tons)
Length: 5.51m (18ft 1in)
Width: 2.54m (8ft 4in)
Height: 2.59m (8ft 6in)
Engine: 111.9kW (150hp) AEC Type 179 6-cylinder petrol
Speed: 26km/h (16mph)
Range: 161km (100 miles)
Armament: 1 x 40mm (1.57in) 2pdr OQF gun, plus 2 x 7.92mm (0.31in) Besa MGs (1 coaxial and 1 ball-mounted in hull front)
Radio: Wireless Set No. 9

Specifications
Crew: 4
Weight: 15.04 tonnes (14.8 tons)
Length: 6.02m (19ft 9in)
Width: 2.59m (8ft 6in)
Height: 2.59 (8ft 6in)
Engine: 253.64kW (340hp) Nuffield Liberty V12 petrol
Speed: 48km/h (30mph)
Range: 145km (90 miles)
Armament: 1 x 40mm (1.57in) 2pdr OQF gun, plus 1 x coaxial 7.92mm (0.31in) Besa MG
Radio: Wireless Set No. 9

▲ **Cruiser Tank Mark IV, A13 Mark II**

Western Desert Force / 7th Armoured Division / 4th Armoured Brigade / 2 RTR

The A13's high speed proved invaluable during Operation *Compass*.

NORTH AFRICA: 1940–43

counter-offensive went on in great secrecy throughout the next few months and were completed by early December.

The detailed plans only covered the initial five-day Operation *Compass* in which 4th Indian and 7th Armoured Divisions were to attack each of the widely separated Italian camps in turn. After the first five days, 4th Indian Division was scheduled to be redeployed to protect the Sudan from Italian forces in Ethiopia and Eritrea, which meant that exploitation of any success would have to be left to 7th Armoured Division and 16th Infantry Brigade.

The first attack by 7 RTR and elements of 4th Indian Division went in against Nibeiwa Camp at dawn on 9 December under cover of a 200-gun barrage. The motorized *Gruppo* Maletti, which formed the camp's garrison, included 23 M11/39s, which were the only Italian medium tanks in the

▲ Cruiser Tank Mark IV, A13 Mark II

Western Desert Force / 7th Armoured Division / 4th Armoured Brigade / 7th Hussars

In common with most other British cruiser and infantry tanks of the period, most A13s were armed with the 2pdr, although a small percentage were fitted with a 76mm (3in) howitzer for the close support role.

Specifications
Crew: 4
Weight: 15.04 tonnes (14.8 tons)
Length: 6.02m (19ft 9in)
Width: 2.59m (8ft 6in)
Height: 2.59 (8ft 6in)
Engine: 253.64kW (340hp) Nuffield Liberty V12 petrol
Speed: 48km/h (30mph)
Range: 145km (90 miles)
Armament: 1 x 40mm (1.57in) 2pdr OQF gun, plus 1 x coaxial 7.92mm (0.31in) Besa MG
Radio: Wireless Set No. 9

▲ Matilda II Infantry Tank

Western Desert Force / 4th Indian Infantry Division / 7 RTR

In their first major actions, Matildas were found to be virtually immune to all Italian weapons, except for heavy artillery. Even direct hits by the 13.6kg (30lb) high explosive shells of 100mm (3.9in) howitzers rarely inflicted serious damage.

Specifications
Crew: 4
Weight: 26.92 tonnes (26.5 tons)
Length: 5.61m (18ft 5in)
Width: 2.59m (8ft 6in)
Height: 2.52m (8ft 3in)
Engine: 2 x 64.8kW (87hp) AEC 6-cylinder diesel
Speed: 13km/h (8mph)
Range: 258km (160 miles)
Armament: 1 x 40mm (1.57in) 2pdr OQF gun, plus 1 x coaxial 7.92mm (0.31in) Besa MG
Radio: Wireless Set No. 11

NORTH AFRICA: 1940–43

Specifications
Crew: 4
Weight: 26.92 tonnes (26.5 tons)
Length: 5.61m (18ft 5in)
Width: 2.59m (8ft 6in)
Height: 2.52m (8ft 3in)
Engine: 2 x 64.8kW (87hp) AEC 6-cylinder diesel
Speed: 13km/h (8mph)
Range: 258km (160 miles)
Armament: 1 x 40mm (1.57in) 2pdr OQF gun, plus 1 x coaxial 7.92mm (0.31in) Besa MG
Radio: Wireless Set No. 11

▼ **Matilda II Infantry Tank**

Western Desert Force / 7th Armoured Division / 7th Armoured Brigade / 4 RTR

This Matilda carries the prominent white-red-white recognition stripes applied to many British AFVs in North Africa.

▲ **Matilda II Infantry Tank**

Western Desert Force / 7th Armoured Division / 4th Armoured Brigade / 2 RTR

A wide variety of desert camouflage schemes were tried at different times – including desert pink. Prolonged exposure to the sun and sandstorms frequently faded camouflage paints, producing strange colours.

Specifications
Crew: 4
Weight: 26.92 tonnes (26.5 tons)
Length: 5.61m (18ft 5in)
Width: 2.59m (8ft 6in)
Height: 2.52m (8ft 3in)
Engine: 2 x 64.8kW (87hp) AEC 6-cylinder diesel
Speed: 13km/h (8mph)
Range: 258km (160 miles)
Armament: 1 x 40mm (1.57in) 2pdr OQF gun, plus 1 x coaxial 7.92mm (0.31in) Besa MG
Radio: Wireless Set No. 11

Specifications
Crew: 4
Weight: 26.92 tonnes (26.5 tons)
Length: 5.61m (18ft 5in)
Width: 2.59m (8ft 6in)
Height: 2.52m (8ft 3in)
Engine: 2 x 64.8kW (87hp) AEC 6-cylinder diesel
Speed: 13km/h (8mph)
Range: 258km (160 miles)
Armament: 1 x 40mm (1.57in) 2pdr OQF gun, plus 1 x coaxial 7.92mm (0.31in) Besa MG
Radio: Wireless Set No. 11

▲ **Matilda II Infantry Tank**

Western Desert Force / 7th Armoured Division / 1st Army Tank Brigade / 7 RTR

This Matilda has a version of the 'radiating lines' camouflage pattern widely used in 1940–41. The lightest colour was primarily applied at the bottom, to lighten shadows around the suspension, with the darkest on top, to lessen the contrast of reflected light.

NORTH AFRICA: 1940–43

▲ **Hunting for the enemy**
A Matilda advances through a minor sandstorm. Poor visibility caused by dust, heat haze and mirages often made accurate gunnery all but impossible.

Italians' morale began to crack, and the death of General Pietro Maletti hastened the final surrender of the 4000 surviving members of the garrison. Attacks on the other camps followed much the same pattern as at Nibeiwa and (after only 48 hours and at the cost of 600 British casualties) their garrisons were destroyed with total losses of 20,000 prisoners, 180 guns and 60 tanks.

This defeat thoroughly demoralized Graziani, who believed that there was no chance of holding the key ports of Bardia and Tobruk against a further British offensive. In his despair he even voiced doubts about the chances of holding the rest of Libya, which undermined his credibility with Mussolini and the *Comando Supremo* in Rome.

front-line. These were surprised and scarcely had time to fire a shot before they were destroyed by the Matildas, which were in any case invulnerable to their feeble 37mm (1.5in) guns.

After dealing with the M11/39s, 7 RTR, with close infantry support, concentrated on overrunning the dug-in Italian artillery. Most gun crews fought to the death, although even hits from the 13.6kg (30lb) shells of the Italians' 100mm (3.9in) howitzers did little more than jam the turrets of one or two Matildas. With the destruction of their guns, the

Bardia and Tobruk

Even as the Western Desert Force drove the Italians back across the Egyptian frontier, O'Connor and Wavell began working on plans for the next stage of the campaign, with the capture of Bardia and Tobruk as the first priority. Given the urgency of the situation, 6th Australian Division began immediate intensive training for the assault on Bardia as soon as it arrived to replace 4th Indian Division. (In a further reorganization, the Western Desert Force was retitled XIII Corps at about this time.)

▲ **Fiat-Ansaldo M11/39**
XIII Corps / 6th Australian Division Cavalry Regiment
The M11/39 was a poorly designed vehicle developed in the late 1930s as a 'breakthrough tank' for the Italian Army. By the time it went into action in 1940, it was already obsolescent and took heavy losses from British 2pdr tank and anti-tank guns. A small number of M11/39s and the later M13/40s were captured in running order and issued to 6th Australian Division Cavalry Regiment to support their Universal Carriers in the assault on Tobruk in January 1941. The large white kangaroo recognition markings were applied to the front, sides and rear of all captured AFVs used by the regiment to minimise the risk of 'friendly fire'.

Specifications
Crew: 3
Weight: 11.175 tonnes (11 tons)
Length: 4.7m (15ft 5in)
Width: 2.2m (7ft 2.5in)
Height: 2.3m (7ft 6.5in)
Engine: 78.225kW (105hp) Fiat SPA 8T V-8 diesel
Speed: 32.2km/h (20mph)
Range: 200km (125 miles)
Armament: 1 x 37mm (1.45in) Vickers-Terni L/40 in limited traverse mount in hull front, plus 2 x 8mm (0.31in) MGs in turret
Radio: None fitted

NORTH AFRICA: 1940–43

The port's garrison included the battered 62nd, 63rd and 64th Divisions, plus the untried 1st and 2nd Blackshirt Divisions. In addition, there were over 100 tanks (mainly L3s, with a handful of M11/39s and the first few M13/40s), plus over 400 guns of all types. These forces held a fortified perimeter complete with anti-tank ditches, wire entanglements and concrete strongpoints stretching for 29km (18 miles) around the harbour.

In theory, the 40,000-strong garrison should easily have been able to hold out for the duration of their one month's water supply, but demoralization had already spread after earlier defeats. The defenders were further shaken by raids of up to 100 RAF bombers during the first nights of 1941. On top of all this, a final blow to Italian morale came in the early hours of 3 January as the battleships *Barham*, *Valiant* and *Warspite* gave impressive naval gunfire support to thicken up the fire of the corps' 160 guns.

The initial barrage concentrated on the area of the defences held by the 1st Blackshirt Division, which was subsequently in no state to put up much resistance to the following Australian assault. Sections of the anti-tank ditch were quickly levelled to allow 7 RTR (which was now down to 23 serviceable Matildas) to support the next phase of the attack. Most Italian positions were only too ready to surrender, but a single counterattack by six M11/39s and M13/40s made some progress before it was broken up by accurate fire from three portee 2pdrs. After this, the attackers were able to work steadily through the defences until the last positions surrendered on 5 January. The majority of the garrison were captured, together with almost all their tanks and guns.

The advance to Bardia had strained O'Connor's supply columns to the limit and made the capture of Tobruk's harbour facilities vital for the next phase of the offensive. The anti-tank ditches, wire and emplacements of the port's perimeter defences, which stretched for almost 48km (30 miles), were held by a total garrison of 25,000 with 200 guns and 87 tanks (including 25 mediums). The old armoured cruiser *San Giorgio* had been damaged in earlier attacks, but was beached in the harbour so that its guns could be used in the defence of the port area.

Tobruk taken

The 7th Armoured Division reached Tobruk on the evening of 6 January, whilst the Australians and 7 RTR followed to complete the process of sealing off the perimeter. Preparations for the attack included the now usual air raids and naval bombardments before the assault went in on the 21st. This used similar tactics to those employed at Bardia, but took into account the lessons learnt in the earlier operation and was influenced by the need to minimize the risks to the invaluable Matildas, only 16 of which could be brought forward in time.

Within an hour of moving off, the Australians had broken through the defences and were widening the breach in the Italian line. Two hours later the attackers were approaching the central strongpoints of Fort Pilastrino and Fort Solaro when they came under fire from a line of 22 dug-in tanks, one of which was destroyed and the rest captured by a daring infantry assault. Counterattacks by other armoured formations were quickly defeated by small groups of Matildas and infantry with artillery support. The failure of these counterattacks triggered the collapse

▶ **Universal Carrier**
Western Desert Force / 6th Australian Division / 16th Australian Infantry Brigade

Almost invariably referred to as the 'Bren Carrier', the Universal Carrier was as indispensable in North Africa as elsewhere for resupply, casualty evacuation and a thousand and one other tasks.

Specifications

Crew: 3	Engine: 63.4kW (85hp) Ford V8 8-cylinder petrol
Weight: 4.06 tonnes (4 tons)	Speed: 52km/h (32mph)
Length: 3.76m (12ft 4in)	Range: 258km (160 miles)
Width: 2.11m (6ft 11in)	Armament: 1 x 14mm (0.55in) Boys anti-tank
Height: 1.63m (5ft 4in)	rifle, plus 1 or 2 x 7.7mm (0.303in) Bren MGs

NORTH AFRICA: 1940–43

of the garrison's morale and most units had surrendered by the evening of the 21st, although those holding the port area did not give in until the next morning. The real importance of the capture of Tobruk lay in its resources, including fuel depots, harbour installations and a water distillation plant that O'Connor desperately needed to supply his planned advance to Benghazi and beyond. The next move was directed against the key crossroads at Mechili, which was held by the *Bambini* Armoured Brigade with a total of 120 M13/40s. Most of these had only just arrived from Benghazi and were scarcely ready for action when the brigade was ordered to attack the leading elements of 7th Armoured Division, which were threatening the desert flank of the main Italian defences covering the coast road at Derna. In their first action on 24 January, a detachment of the brigade's M13/40s forced the 7th

Specifications
Crew: 4
Weight: 4.26 tonnes (4.2 tons)
Length: 4.78m (15ft 8in)
Width: 2.06m (6ft 9in)
Height: 2.21m (7ft 3in)
Engine: 52kW (70hp) Morris Commercial
4-cylinder petrol
Speed: 72km/h (45mph)
Range: 386km (240 miles)
Armament: 1 x 14mm (0.55in) Boys anti-tank rifle, plus 1 x 7.7mm (0.303in) Bren MG
Radio: Wireless Set No. 9

▲ **Morris CS9 Armoured Car**
Western Desert Force / 7th Armoured Division / 7th Armoured Brigade / 11th Hussars
The 11th Hussars were equipped with 30 of these armoured cars in the early months of the desert war. Although their armament of a Boys anti-tank rifle and Bren gun was feeble, they were popular for their ability to cross soft sand.

▲ **Marmon-Herrington Armoured Car Mark II**
Eighth Army / 2nd Armoured Division / 3rd Armoured Brigade / King's Dragoon Guards
Marmon-Herrington armoured cars were built in South Africa using a variety of imported components. Almost 900 Mark IIs were produced, the majority of which were sent to North Africa.

Specifications
Crew: 4
Weight: 6.096 tonnes (6 tons)
Length: 5.18m (17ft)
Width: 2m (6ft 6in)
Height: 2.67m (8ft 9in)
Engine: 63kW (85hp) Ford V8 petrol
Speed: 80km/h (50mph)
Range: 322km (200 miles)
Armament: 1 x 14mm (0.55in) Boys anti-tank rifle, plus 2 x 7.7mm (0.303in) Bren MGs (1 in the turret and 1 AA)
Radio: Wireless Set No. 9

NORTH AFRICA: 1940–43

Hussars' light tanks to withdraw after knocking out two of their three cruisers. The pursuing Italian armour was ambushed by a squadron of 2 RTR's cruisers which, firing from hull-down positions, destroyed seven M13/40s in as many minutes. This action, combined with the usual greatly exaggerated reports of British strength, prompted the withdrawal of the *Bambini* Armoured Brigade before 7th Armoured Division could completely surround their positions. The way was now clear for probing attacks against the desert flank of the Derna position, which was evacuated during the night of 28/29 January.

As the Italians withdrew westwards along the coast road, O'Connor planned a 240km (150-mile) advance across the desert by 7th Armoured Division to cut off their retreat near Beda Fomm. Whilst this was going on, the Australians were to keep up the pressure of the pursuit along the coast road to

▲ Marmon-Herrington Armoured Car Mark II
Eighth Army / 7th Armoured Division / 4th South African Armoured Car Regiment

The Marmon-Herrington's small turret with a standard armament of a Boys anti-tank rifle and a Bren, plus one or two AA machine guns, was frequently up-gunned with heavier weapons such as captured 20mm (0.79in) AA guns.

Specifications
Crew: 4
Weight: 6.096 tonnes (6 tons)
Length: 5.18m (17ft)
Width: 2m (6ft 6in)
Height: 2.67m (8ft 9in)
Engine: 63kW (85hp) Ford V8 petrol
Speed: 80km/h (50mph)
Range: 322km (200 miles)
Armament: 1 x 14mm (0.55in) Boys anti-tank rifle, plus 2 x 7.7mm (0.303in) Bren MGs (1 in the turret and 1 AA)
Radio: Wireless Set No. 9

Specifications
Crew: 1
Weight: 3.05 tonnes (3 tons)
Length: 6.58m (21ft 7in)
Width: 2.49m (8ft 2in)
Height: 3m (9ft 9in)
Engine: 71kW (95hp) Ford V8 petrol
Speed: 80km/h (50mph)
Range: 274km (170 miles)

▲ Chevrolet WA
Eighth Army / Long Range Desert Group

This Chevrolet is typical of the wide range of 'soft-skinned' vehicles used by the Long Range Desert Group for raiding and reconnaissance missions far behind enemy lines. A vast assortment of weaponry was carried, ranging from machine guns to anti-tank and light AA guns.

NORTH AFRICA: 1940–43

▼ B Squadron, 2nd Royal Tank Regiment, 1940

During the first months of the desert war, 2 RTR operated a mixed bag of A9, A10 and A13 cruiser tanks. All three types had differing levels of mobility and armour protection, but were nonetheless used very effectively against far larger Italian forces.

convince Graziani that this was still the main threat. The spearhead of the dash for Beda Fomm was a small group equipped with wheeled vehicles and designated 'Combeforce' after its commander, Lieutenant-Colonel John Combe. It had a total strength of only 2000 men and comprised: 11th Hussars with Rolls Royce and Morris armoured cars; 2nd Battalion, The Rifle Brigade; C Battery, 4th Royal Horse Artillery, with six 25pdr guns; and 106th Battery, Royal Artillery, with nine 37mm (1.5in) Bofors portee anti-tank guns.

Combeforce in position

By midday on 5 February, Combeforce had set up an ambush on the coast road at Sidi Saleh, a few kilometres south-west of Beda Fomm. They were just a couple of hours ahead of the retreating Italians, who made a series of increasingly desperate but uncoordinated attacks in their attempts to break through. Initially these attacks were small-scale affairs as of all the varied units heading the retreat only the 10th *Bersaglieri* were first-line combat troops and they were without any armoured support. Inevitably, the pressure against Combeforce began to mount as more Italian units arrived, and the situation was becoming serious when the leading elements of 4th Armoured Brigade arrived late that afternoon. These were directed against the flanks of the steadily lengthening enemy column, shooting up seemingly endless lines of halted soft-skinned vehicles and taking 800 prisoners. The most valuable prizes were petrol tankers, which were immediately used to refuel the brigade's tanks, several of which had run dry on the battlefield itself.

That night, both sides gathered their tank strengths for the coming battle. The *Bambini* Armoured Brigade was ordered to detach 60 of its M13/40s from the rearguard to help batter a way

NORTH AFRICA: 1940–43

> **CAMOUFLAGE & MARKINGS, 1940–41**
>
>
>
> The most common British camouflage was an overall sand colour, occasionally over-painted with a disruptive pattern of slate grey (top). In 1937, the 11th Hussars' armoured cars were repainted in a disruptive scheme of sand and deep red, which may still have been in use in 1940–41.
>
>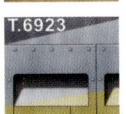
>
> Shortly after the desert war began, a new paint scheme was adopted by 1 RTR and 6 RTR, plus some other units. This consisted of radiating diagonal bands of (from top to bottom) slate grey, light grey and sand.
>
> British tactical markings (bottom) were unchanged from those described elsewhere, but photographs seem to indicate that they were rarely used in the front-line.

combined efforts accounted for an estimated total of 79 M13/40s.

By the end of the day's fighting, the Italians were down to barely 30 serviceable tanks, all of which were committed to a final desperate attack at first light on 7 February. This was launched with the support of all the remaining artillery and came close to breaking through Combeforce's positions before being destroyed by accurate fire from 106th Battery's guns. Italian morale collapsed with this defeat and the mass surrenders began, the final total for this action reaching 25,000 prisoners, plus over 100 tanks, 216 guns and 1500 other vehicles.

No further advance

O'Connor pressed for the exploitation of the victory, proposing a further advance to occupy the whole of Libya, which would have been a massive blow to Mussolini's prestige and might well have forced Italy out of the war. He was overruled by political pressures to divert forces to support Greece in its counter-offensive against the Italian forces that had invaded from Albania. Despite this, the campaign had ended in a stunning British success, amply illustrated by a comparison of the losses on each side:

- British: 500 dead, 55 missing and 1373 wounded.
- Italian: 10 divisions destroyed (130,000 prisoners); 180 medium tanks and over 200 light tanks captured; 845 guns captured.

through with support from all the available artillery. The British also frantically worked to raise the number of AFVs fit for action, which eventually totalled 32 cruisers and 53 light tanks plus some armoured cars.

Throughout the next day, a succession of Italian attacks were beaten off, notably at the small hill known as the Pimple, where accurate fire from nine hull-down A13s of A Squadron, 2 RTR, broke up two assaults, each led by at least 20 M13/40s. The A13s were running critically short of 2pdr ammunition by the time the slower A9s and A10s of C Squadron were able to join the battle, but their

El Agheila to El Alamein
1941–42

The string of victories that culminated at Beda Fomm had made British commanders dangerously overconfident. Matters were made worse by the transfer of battle-hardened formations to Greece and their replacement by inexperienced units that would inevitably take time to adjust to the demands of desert warfare.

THE EASY VICTORIES of the opening months of the desert war had left the impression that mobility and surprise were guarantees of success, and conventional formations tended to be broken up to form large numbers of small raiding units, each with their own detachments of armour, artillery and infantry. Although some of these small units were very effective, they became used to fighting their own 'private wars' and lost the knack of efficient cooperation with other formations in larger actions. The breaking-up of the sophisticated artillery command structure was perhaps the worst single

NORTH AFRICA: 1940–43

effect of this process. As a result, intense, concentrated bombardments could rarely be laid when they were really needed, most notably to counter Rommel's expertly sited anti-tank batteries, which inflicted such heavy losses on British armour. Dug-in 88s (3.5in weapons) and 50mm (2in) Pak 38s gave bloody reminders of lessons that should have been learnt after Arras. In the early stages of Operation *Battleaxe*, the towed artillery assigned to support 4 RTR's Matildas bogged down in soft sand, leaving the tanks to be picked off by 88mm (3.5in) weapons dug-in on Halfaya Ridge.

▲ **Infantry Tank Mark III, Valentine Mark II**
Eighth Army / XIII Corps / 1st Army Tank Brigade
Valentines were introduced to supplement the Matildas in 1941 and first went into action with 8 RTR during Operation *Crusader* in November of that year. During 1942 they steadily replaced Matildas in the infantry tank regiments.

Specifications
Crew: 3
Weight: 17.27 tonnes (17 tons)
Length: 5.89m (19ft 4in)
Width: 2.64m (8ft 8in)
Height: 2.29m (7ft 6in)
Engine: 97.73kW (131hp) AEC 6-cylinder diesel
Speed: 24km/h (14.9mph)
Range: 145km (90 miles)
Armament: 1 x 40mm (1.57in) 2pdr OQF gun, plus 1 x coaxial 7.92mm (0.31in) Besa MG
Radio: Wireless Set No. 11

▲ **Infantry Tank Mark III, Valentine Mark II**
Eighth Army / XIII Corps / 1st Army Tank Brigade
The Valentine was greatly appreciated for its mechanical reliability, which was far higher than that of most contemporary British tanks.

Specifications
Crew: 3
Weight: 17.27 tonnes (17 tons)
Length: 5.89m (19ft 4in)
Width: 2.64m (8ft 8in)
Height: 2.29m (7ft 6in)
Engine: 97.73kW (131hp) AEC 6-cylinder diesel
Speed: 24km/h (15mph)
Range: 145km (90 miles)
Armament: 1 x 40mm (1.57in) 2pdr OQF gun, plus 1 x coaxial 7.92mm (0.31in) Besa MG
Radio: Wireless Set No. 11

NORTH AFRICA: 1940–43

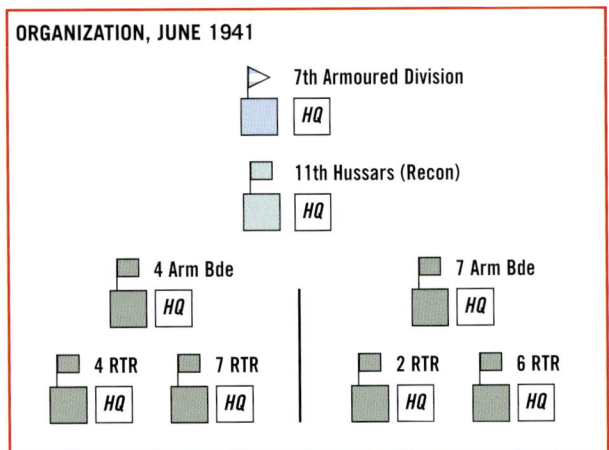

ORGANIZATION, JUNE 1941

These problems were worsened by the totally inadequate provision for HE-firing weapons in British tank designs; it was confined to the small number of close support (CS) versions of each type, which were armed with the 76mm (3in) CS howitzer. (In 1941 only six of the 52 cruiser tanks in each armoured regiment were CS variants.) Not only were the numbers of CS tanks too low, but the weapon itself was unimpressive, with a badly designed HE round of poor lethality that had a maximum range of no more than 2200m (2407 yards). (Contrary to popular belief, there was a 2pdr HE shell, but very little has come to light on this ammunition, which seems only to have been issued to AFVs in the Far East from 1943 onwards.)

All too often these factors resulted in British tanks being committed to costly frontal charges in desperate attempts to get within machine-gun range of the enemy anti-tank guns.

No real solution was found until the US-supplied Grants and Lees began to enter service in May 1942, armed with 75mm (2.9in) M2 and M3 guns firing a good HE shell (much more effective than that of the 76mm/3in CS howitzer) out to 12,000m (13,130 yards). The M2 and M3 also had a respectable anti-tank performance for the time, although the limited traverse sponson mounting made it impossible to fire from hull-down positions.

The same 75mm (2.9in) M3 weapon, but now in a conventional turret, was the main armament of the Shermans which began to be issued a few months later. This layout solved most of the tactical problems posed by the earlier sponson mountings, and long-range indirect HE fire was made more effective by the fitting of the Azimuth Indicator M19.

Enter the Desert Fox

Rommel arrived in North Africa on 12 February 1941 and quickly appreciated the urgency of boosting the Italians' morale, which had been badly shaken by their defeat at Beda Fomm. The first probing attacks were launched as early as 24 February and were followed up by a full-blown offensive on 24 March when the German 5th Light Division had completed its concentration.

▲ **Heat and dust**
A column of Crusaders kicks up dust. Vehicles could rarely move in the desert without creating large clouds of dust that could be seen for kilometres.

NORTH AFRICA: 1940–43

This timing totally surprised Rommel's opponents, who had assumed that no major operations would be launched until the Germans had taken a month or so to acclimatize and settle in. Aided by the inexperience of so many Allied front-line units, this surprise attack achieved a striking success, including the capture of General O'Connor, who had masterminded the victorious Beda Fomm campaign. After little more than a month, Axis forces had recaptured almost all the territory lost during the previous winter and were again on the Egyptian frontier. Only the fortified port of Tobruk, which Rommel desperately needed to ease his acute supply problems, stubbornly held out against all attacks.

Stung by the dramatic Axis advance, General Wavell in Cairo planned a counterattack, codenamed Operation *Brevity*, which opened on 15 May. The objective was the recapture of the vital Halfaya Pass together with Sollum and Fort Capuzzo as a preliminary to raising the siege of Tobruk. The pass and Fort Capuzzo were seized, but Rommel reacted quickly and retook both by 27 May.

By 15 June, Wavell had been reinforced and launched Operation *Battleaxe* with a total of almost 400 tanks, again aimed at breaking through to Tobruk before driving westwards to secure a line from Derna to Mechili. Although British reading of the German 'Enigma' codes provided useful information, Rommel's own expert signals intelligence staff were able to counter this through their intercepts of many of Wavell's orders. With neither side having a clear advantage in the 'intelligence war', superior German tactical ability proved to be the decisive factor – especially their expertise in deploying anti-tank batteries, which had a distinct edge over AFVs provided that the gun crews had time to measure ranges and set up range markers. (This was largely due to the special conditions of desert warfare in which mirages, heat shimmer and dust clouds made it very difficult to identify targets and even harder to accurately judge ranges.)

▲ **Get a move on!**
Crews race to their Stuart Is, autumn 1941. They all wear early pattern US tankers' helmets, which were supplied with Lend-Lease AFVs in 1941–42.

▲ **Cruiser Tank Mark VI, Crusader Mark I**
Eighth Army / 7th Armoured Division / HQ / 22nd Armoured Brigade
Crusaders began to replace the earlier cruiser tanks in North Africa from mid-1941. A9s were phased out of service at about this time, but A10s and A13s soldiered on until the end of the year.

Specifications
Crew: 5
Weight: 19.26 tonnes (18.95 tons)
Length: 5.99m (19ft 8in)
Width: 2.64m (8ft 8in)
Height: 2.24m (7ft 4in)
Engine: 253.64kW (340hp) Nuffield Liberty Mark III V12 petrol
Speed: 44km/h (27mph)
Range: 161km (100 miles)
Armament: 1 x 40mm (1.57in) 2pdr OQF gun, plus 2 x 7.92mm (0.31in) Besa MGs (1 coaxial and 1 in sub-turret)
Radio: Wireless Set No. 11

NORTH AFRICA: 1940–43

In tank-versus-tank actions, the British and Germans were quite evenly matched, although 2pdr shot tended to shatter against the Panzers' face-hardened armour at close range, a fault which was not cured until capped shot (APCBC, or armour-piercing cap ballistic cap) was issued in May 1942. After making some initial gains, the British offensive was fought to a standstill within a couple of days and their outmanoeuvred armour only just escaped being surrounded. The respective losses (220 British tanks lost, of which 87 were complete write-offs, against 25 Panzers destroyed) clearly showed how well Rommel had done with his outnumbered forces and marked the beginning of his reputation as the 'Desert Fox'.

Tobruk campaign

In the aftermath of the failure of Brevity and Battleaxe, Wavell was posted to India and replaced by General Claude Auchinleck as commander-in-chief in Cairo, whilst General Alan Cunningham took over command of the forces in the desert, which were now reorganized as the Eighth Army. Rommel's victories were recognized by his promotion to *General der Panzertruppe* (lieutenant-general) on 1 July and his command was redesignated *Panzergruppe Afrika* on 15 August. The first operation of the *Panzergruppe* (in September 1941) was a raid on what was believed to be a major British supply dump to build up stocks of fuel before renewing the assault on Tobruk, which was planned for 23 November. This raid was a failure with the loss of 30 AFVs and the delay allowed Auchinleck to launch his own offensive (Operation *Crusader*) on 18 November with over 750 tanks, including 280 of the newly supplied M3 Stuarts, supported by 600 guns.

Once again, the objective was to raise the siege of Tobruk and the opening moves went well – the major airfield at Sidi Rezegh was captured and 19 Axis aircraft were destroyed on the ground. The inevitable German counterattacks began the next day, with each of the two Panzer divisions massing all available resources to strike concentrated blows at the scattered British brigades. In four days of fighting two of these brigades lost some 300 tanks between them. As Rommel was later to remark to a captured British officer: 'What does it matter if you have two tanks to my one, when you spread them out and let me smash them in detail?'

New broom

General Cunningham wanted to withdraw, but was replaced by General Neil Ritchie, who was under strict orders to continue with the offensive. Rommel was convinced that the Eighth Army would indeed be forced to fall back and he led the bulk of his forces in a dash for the Egyptian frontier to cut off its line of retreat. This spectacular advance caused a temporary panic in Egypt, but by abandoning the battlefield around Sidi Rezegh, Rommel took the pressure off the battered British armoured units, which were allowed time to recover and repair many of their damaged tanks whilst New Zealand forces broke through to the Tobruk garrison.

By the time that the *Deutsches Afrikakorps* (DAK) was able to regroup, it was down to 40 battleworthy tanks whilst the Italians were even worse off with only

▶ M3 Stuart I Light Tank

Eighth Army / XXX Corps / 7th Armoured Division / 4th Armoured Brigade / 8th Hussars

Eighty-four Stuarts arrived in North Africa in July 1941 and by November the total had risen to 280, which equipped all three battalions of 4th Armoured Brigade.

Specifications

Crew: 4
Weight: 12.7 tonnes (12.5 tons)
Length: 4.53m (14ft 10in)
Width: 2.24m (7ft 4in)
Height: 2.64m (8ft 8in)
Engine: 186.25kW (250hp) Continental W-670-9A 7-cylinder petrol
Speed: 58km/h (36mph)
Range: 110km (70 miles)
Armament: 1 x 37mm (1.5in) M5 gun, plus 3 x 7.62mm (0.3in) MGs (1 AA, 1 coaxial and 1 ball-mounted in hull front)
Radio: SCR210

NORTH AFRICA: 1940–43

30 serviceable tanks. However, it was the chronic supply problems that posed the greatest threat to the Axis cause in the theatre as virtually two-thirds of the 122,000 tonnes (120,000 tons) of stores sent to North Africa in recent weeks had been sunk en route. In early December a determined British attack convinced Rommel that he had to retreat, and by Christmas he was back where he had started 10 months earlier.

The next round in the desert war opened at the beginning of 1942 with the long-awaited arrival of reinforcements for the DAK, including 74 AFVs, which allowed Rommel to launch his next offensive on 21 January. Within a couple of weeks this pushed the front-line almost 400km (250 miles) eastwards to the fortified Gazala Line covering Tobruk, where it

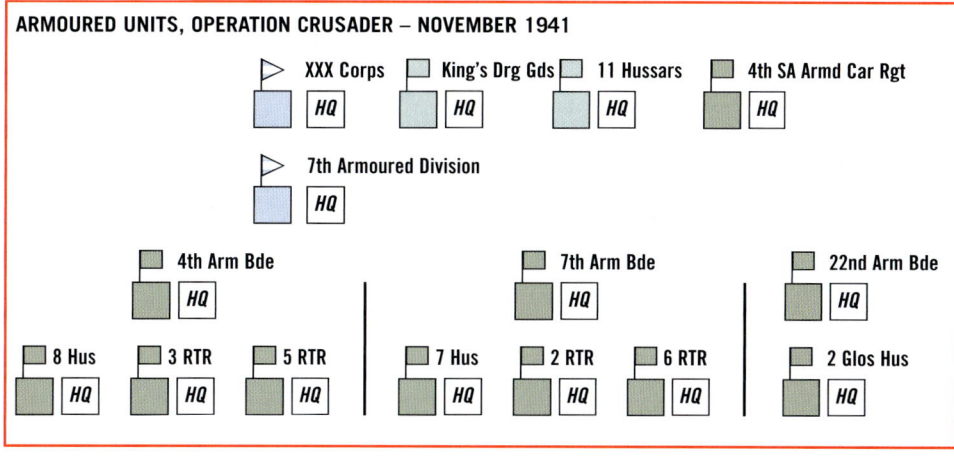

▼ C Squadron, 8th Hussars

The Stuart was highly popular with its crews for its reliability and speed. It was widely referred to as the 'Honey', supposedly after a driver remarked 'She's a honey!' on returning from his first test drive.

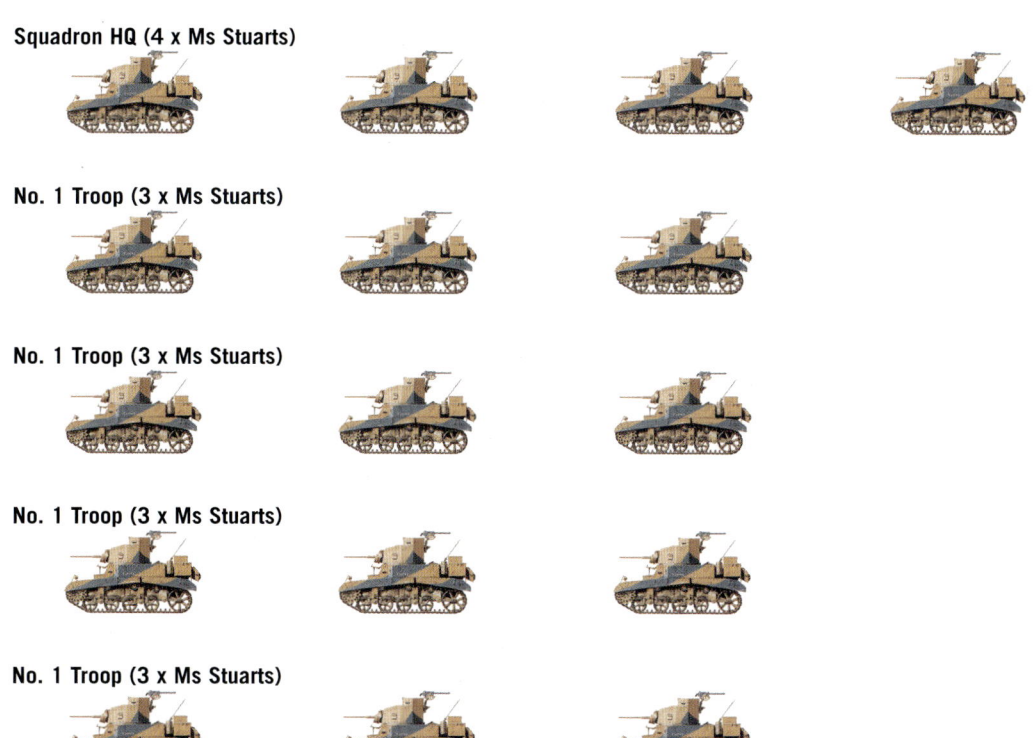

NORTH AFRICA: 1940–43

remained for the next four months as both sides built up their strength for the next move.

By the end of May, the Axis forces were heavily outnumbered in terms of tanks, although they had a distinct edge in numbers of aircraft. Despite the obvious difficulties posed by Allied tank strengths, which for the first time included the Grant, Rommel considered that his forces still had sufficient tactical superiority to ensure victory and attacked again on 26 May, only just before Ritchie was ready to launch his own offensive. Whilst Italian infantry divisions attacked the coastal end of the Gazala Line, the DAK plus the *Ariete* Armoured Division moved to outflank the southern end of the Allied line, held by the Free French at Bir Hacheim. Although the Bir Hacheim garrison beat off all assaults, the main Axis armoured thrust was able to push into the Eighth Army's rear areas before any major counterattacks developed. These were determined efforts which, for a few critical days, trapped Rommel against the Allied minefields in an area which became known as the Cauldron. Eventually the pressure was eased by the defeat of a major Allied assault followed by the evacuation of Bir Hacheim on 10/11 June.

Barely 24 hours later, the DAK broke out of the Cauldron and raced for the Tobruk perimeter, reimposing the siege on 18 June. This time, however, the port's garrison was in no condition to hold out for more than a few days and, on 21 June, it surrendered to a jubilant Rommel, who was promoted to field-marshal in recognition of his victory, having been in the rank of full general for less than six months.

In the aftermath of defeat, command of the Eighth Army changed again as Auchinleck dismissed Ritchie and took over in person. Allied forces fell back on Mersa Matruh before continuing their retreat to El Alamein, the final defensible line covering the Nile Delta. The DAK was in hot pursuit and had closed up to the El Alamein positions by 1 July, but failed to break through in a series of attacks launched over the next few days.

> **CAMOUFLAGE & MARKINGS, 1942**
>
>
>
> In general, there were only minor changes to camouflage and markings from those described in the preceding section. Most British AFVs were simply painted in overall sand yellow or sand brown (top), but no matter what the nationality, the actual colours could vary greatly after prolonged exposure to the fierce sun and an occasional sandstorm. A few units used more elaborate camouflage schemes, adding disruptive patterns of darker brown, green, grey or black to the original sand finish (bottom).

▲ **M3 Lee Medium Tank**

Eighth Army / 1st Armoured Brigade / 1 RTR

The Grants-M3s modified to meet British specifications by the adoption of a new low turret without a cupola were first used in action at Gazala in May 1942 when a total of 167 equipped three armoured brigades. Standard M3s, designated Lees in British service, were also sent to North Africa to replace combat losses and a total of 210 Grants/Lees were in service by October 1942.

Specifications

Crew: 7
Weight: 27.22 tonnes (26.7 tons)
Length: 5.64m (18ft 6in)
Width: 2.72m (8ft 11in)
Height: 3.12m (10ft 3in)
Engine: 253.5kW (340hp) Continental R-975-EC2 radial petrol
Speed: 42km/h (26mph)
Range: 193km (120 miles)
Armament: 1 x 75mm (2.9in) M2 or M3 gun, 1 x 37mm (1.5in) M4 or M5 gun, plus 4 x 7.62mm (0.3in) MGs (1 in commander's cupola, 1 coaxial and 2 fixed forward-firing)
Radio: SCR508

NORTH AFRICA: 1940–43

▲ **Covering fire**
A Daimler Mk I armoured car provides fire support for a British infantry attack, October 1942.

▼ **'Bombing-up'**
A Sherman's crew stow ammunition for the tank's 75mm (2.9in) main armament. Early Shermans carried 90 rounds of 75mm (2.9in) ammunition; later models stowed 97 rounds.

Allied counterattacks later in July had only limited success, and stalemate set in with neither side able to achieve a decisive result. This led to yet another change in command of the Allied forces, with Auchinleck replaced by General Harold Alexander whilst General Bernard Montgomery took over the Eighth Army. His efforts to rebuild its self-confidence were greatly helped by the stream of new AFVs arriving to re-equip its armoured formations. Rommel realized that he could not hope to match the scale of the Allied build-up and decided to attack on 30 August before the odds became too great.

Failed attack

The plan was similar to that used at Gazala, with holding attacks launched near the coast, whilst the main effort was aimed at outflanking the southern end of the Allied line. Once this was achieved, the attack would swing north-eastwards to take the Alam el Halfa ridge, cutting the Allied supply lines and unhinging the entire El Alamein position. Even before the offensive began, there were disturbing signs of increased Allied strength – Axis armour was heavily bombed as it began to concentrate in the days leading up to the attack.

The early stages of the advance were slowed by the extensive minefields and once through these, the DAK found that Alam el Halfa was held in strength

by two armoured brigades backed up by ample artillery and air support. After two days of fruitless attacks, Rommel admitted that a breakthrough was impossible and pulled back to his start line, with the loss of more than 50 tanks. British losses totalled almost 70 tanks, but, unlike Rommel's, these could easily be replaced and the balance of forces had now swung decisively in the Allies' favour.

NORTH AFRICA: 1940–43

El Alamein to Tunis
1942–43

The defensive victories at the First Battle of El Alamein and Alam el Halfa had boosted confidence throughout the Eighth Army and had created an atmosphere in which Montgomery's reorganization and retraining were readily accepted.

THE FORMATION ONCE again became a key battlefield player, curbing the tendency to deploy armour in small units each of which could be defeated in succession by properly concentrated Panzer divisions. Most importantly, Montgomery emphasized training in all-arms cooperation, in which armour, infantry and artillery worked together as battlefield teams. A steadily increasing flow of new AFVs, notably Shermans and Priests plus the first Matilda Scorpion flail tanks, provided the essential equipment for these teams to operate effectively and gave them a decisive advantage over their opponents.

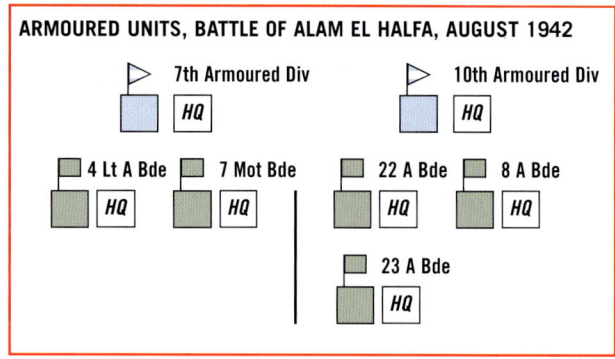

Axis armour 1942–43

This period saw the introduction of a new generation of German AFVs, including up-gunned Stugs, Panzer IVs and the first Tiger Is. A generous allocation of these AFVs to the DAK might have partially offset Allied numerical superiority, but even Rommel could not compete with the overriding needs of the Eastern Front until the prospect of the complete loss of North Africa prompted panic measures in a futile attempt to stave off defeat.

If German problems were bad, they were trivial compared with those confronting Italian armoured units, whose equipment was increasingly outclassed

▲ **Cruiser Tank Mark VI, Crusader Mark III**
Eighth Army / XXX Corps / HQ 23rd Armoured Brigade
The first towed 6pdr anti-tank guns were issued to front-line units in May 1942, but the Crusader III, which carried the 6pdr, was not ready until August of that year. Despite the 6pdr's anti-tank performance being better than that of the US 75mm (2.9in) in the Lee/Grant and Sherman, it was not as popular, because the American gun's superior HE shell was far more useful in dealing with the enemy anti-tank batteries, which posed the greatest threat at that time. Moreover, 6pdr HE was rarely issued and its very existence was often forgotten.

Specifications
Crew: 3
Weight: 20.07 tonnes (19.75 tons)
Length: 6.3m (20ft 8in)
Width: 2.64m (8ft 8in)
Height: 2.24m (7ft 4in)
Engine: 253.64kW (340hp) Nuffield Liberty Mark IV V12 petrol
Speed: 44km/h (27mph)
Range: 161km (100 miles)
Armament: 1 x 57mm (2.24in) 6pdr OQF gun, plus 1 coaxial 7.92mm (0.31in) Besa MG
Radio: Wireless Set No. 11

by Allied AFVs. The various types of Semovente self-propelled guns were useful assault weapons, but the majority mounted short 75mm (2.9in) L/18 guns that had a poor anti-tank performance. Deliveries of more powerfully armed AFVs were slow – barely 30 of the potent Semovente 90/53 were available by late 1942 and several promising types had only just entered production by the time that Italy surrendered in September 1943.

El Alamein

The failure of his offensives at the First Battle of El Alamein and at Alam el Halfa in July/August 1942 left Rommel with seriously weakened armoured forces. His losses, totalling more than 50 tanks, were especially grave given the slow trickle of replacements along a lengthy and dangerously exposed supply line that was being subjected to increasingly heavy air attacks. One of the most damaging effects of these interdiction operations was to cause a chronic fuel shortage throughout the Axis forces, which severely limited their freedom of manoeuvre.

Rommel's priority was now to establish strong defence lines in readiness for the offensive that Montgomery was certain to launch once his build-up was complete. Wherever possible, Allied minefields in captured sectors were used to form part of the Axis defences. These were thickened and extended until the 65km (41-mile) front from the Mediterranean to the virtually impassable terrain of the Qattara Depression was clogged with 500,000 mines. Behind the minefields lay defence lines up to 3km (1.9 miles) deep, held by infantry and artillery (and with the Italian formations bolstered by flanking German units). Axis armour waited in reserve, ready to counterattack as soon as the Allied offensive was blunted by the minefields and fortified positions.

By late October 1942, Montgomery's preparations were complete and he launched the Second Battle of El Alamein on the 23rd. His offensive began at 21:40 with a 30-minute bombardment by 1000 guns and a series of air attacks before the assaults went in, directed against the northern and southern ends of the Axis line.

Eighth Army AFV Units, Oct 1942	Constituent Units	Commanding Officer	Vehicles	Strength
British 7th Armoured Div	1st Household Cav Rgt 11th Hussars	Maj-Gen John Harding		
– 4th Light Armoured Bde	Royal Scots Greys 4th/8th Hussars	Brig Marcus G. Roddick	Stuart Grant	57 14
– 22nd Armoured Bde	1st Royal Tank Regiment 5th Royal Tank Regiment	Brig George 'Pip' Roberts	Grant Crusader Stuart	57 46 19
British 1st Armoured Div	12th Royal Lancers 4th/6th South African Armd Car Rgt	Maj-Gen Raymond Briggs		
– 2nd Armoured Bde	2nd Dragoon Guards (Queen's Bays) 9th Queen's Royal Lancers 10th Royal Hussars Yorkshire Dragoons	Brig Arthur Fisher	Sherman Crusader	92 68
– 7th Motor Bde	2nd Btn King's Royal Rifle Corps 2nd Btn Rifle Brigade 7th Btn Rifle Brigade	Brig Thomas J. Bosville		
British 10th Armoured Div	Royal Dragoons	Maj-Gen Alexander Gatehouse		
– 8th Armoured Bde	3rd Royal Tank Regiment The Nottinghamshire Yeomanry Staffordshire Yeomanry	Brig Edward C.N. Custance	Crusader Grant Sherman	45 57 31
– 24th Armoured Bde	41st Royal Tank Regiment 45th Royal Tank Regiment 47th Royal Tank Regiment	Brig Arthur G. Kenchington	Sherman Crusader	93 45

NORTH AFRICA: 1940–43

Rommel was on sick leave in Germany when the battle began, thoroughly exhausted by 18 months of intensive operations in gruelling desert conditions. He had handed over to General Georg Stumme, who suffered a fatal heart attack just as the battle began, but despite this, the DAK put up a fierce fight over 11 days, inflicting losses of up to 90 per cent on some Allied armoured units.

Rommel's return to the front to resume command on 25 October boosted morale, but could do nothing to alter the two-to-one Allied superiority in AFVs. Over 1000 Allied tanks, including 250 Shermans, faced barely 500 Axis tanks, of which more than half were obsolescent Italian vehicles. This numerical superiority meant that the Allies could accept heavy losses in exchange for the destruction of the DAK's armour – on 24/25 October 15th Panzer Division's counterattack was halted with the loss of all but 31 of its 119 tanks.

Torch landings

Sheer weight of enemy numbers coupled with ever worsening fuel shortages eventually forced Rommel to acknowledge that, despite 'stand fast' orders from Hitler and Mussolini, a retreat was unavoidable if he was going to save any of his units. The withdrawal began on 4 November, and four days later the Allies began Operation *Torch* – major landings in Morocco and Algeria (then French North Africa) aimed at Rommel's base areas, particularly the key ports of Tunis and Bizerta.

No one was certain how the Vichy French garrisons would react to this invasion, and American forces led the first wave in the hope of avoiding open fighting. There was some resistance to the first landings, but within 48 hours a ceasefire had been negotiated and the route to Tunis seemed to be open. The deadly threat posed by Operation Torch was immediately apparent to the Germans, who reacted swiftly, flying in reinforcements to protect their Tunisian bases and establishing a 15,000-strong

> **CAMOUFLAGE & MARKINGS**
>
>
> There was very little change to camouflage schemes until the fighting spread to Tunisia, where there was more vegetation, especially during the wetter winter months. Both sides began to make more use of camouflage schemes, which included at least some greenish paint – a few of the Tiger Is encountered may well have been painted in overall olive green as were some British and most US AFVs during the final stages of the North Africa campaign.

▲ **Dominating the desert**
The Sherman represented a major advance in American AFV design, as it was the first US tank with a turret-mounted 75mm (2.9in) gun. Deliveries to North Africa began in August 1942 and a total of 270 had arrived by October of that year.

NORTH AFRICA: 1940–43

garrison (with 130 AFVs) by the end of November. This proved to be sufficient to halt the First Army's advance from French North Africa, which had made alarming progress – on 27 November Stuarts of the US 1st Tank Battalion and the Derbyshire Yeomanry's Daimler armoured cars raided Djedeida airfield, less than 15km (9 miles) from Tunis.

German counterattack

By early December, General Hans-Jürgen von Arnim, commanding the German forces in Tunisia, was ready to launch his own offensive against the opposing Allied units, which were by now seriously overstretched. Heavy winter rains slowed operations, but affected the inexperienced American forces far more than their veteran German opponents, most notably on 10 December when Combat Command B (CCB), 1st US Armored Division, panicked during a night withdrawal and drove off wildly, eventually abandoning a total of 18 tanks, 41 guns and 130 other vehicles, all bogged down in thick mud. Further German spoiling attacks in January 1943 imposed more delays on the Allied build-up in Tunisia and confirmed the dangerous amateurishness of the American command and control structure.

Whilst these events were unfolding, Rommel, away to the east, was conducting an expertly handled withdrawal westwards towards Tunisia, imposing temporary halts on the Eighth Army's pursuit at El Agheila (23 November–13 December) and at Buerat (26 December–15 January). The Buerat Line was significantly weakened by the transfer of 21st

▲ Infantry Tank Mark IV, Churchill Mark IV
First Army / 4th Infantry Division / 21st Tank Brigade

A trials unit fielded six Churchill IIIs in the Second Battle of El Alamein, but it was not until the Tunisian campaign that the Mark IV was first used in any numbers, sometimes serving alongside Churchill Is in the same units.

Specifications

Weight: 39.62 tonnes (39 tons)
Length: 7.44m (24ft 5in)
Width: 2.74m (9ft)
Height: 3.25m (10ft 8in)
Engine: 261.1kW (350hp) Bedford 12-cylinder petrol
Speed: 25km/h (15.5mph)
Range: 193km (120 miles)
Armament: 1 x 57mm (2.24in) 6pdr OQF gun, plus 2 x 7.92mm (0.31in) Besa MGs (1 coaxial and 1 ball-mounted in hull front)
Radio: Wireless Set No. 19

▶ Humber Armoured Car Mark II
Eighth Army / 7th Armoured Division / 11th Hussars

The Humber Mark II was a 4 x 4 vehicle armed with a 15mm (0.59in) Besa heavy machine gun and a coaxial 7.92mm (0.31in) Besa machine gun. A total of 440 vehicles were produced, many of which saw service in North Africa.

Specifications

Crew: 3
Weight: 7.213 tonnes (7.1 tons)
Length: 4.57m (15ft)
Width: 2.18m (7ft 2in)
Height: 2.34m (7ft 10in)
Engine: 67kW (95hp) Rootes 6-cylinder petrol
Speed: 72km/h (45mph)
Range: 402km (250 miles)
Armament: 1 x 15mm (0.59in) Besa HMG, plus 1 x coaxial 7.92mm (0.31in) Besa MG
Radio: Wireless Set No. 19

NORTH AFRICA: 1940–43

Panzer Division to reinforce von Arnim, and when the position was outflanked by Montgomery's next attack, the DAK was pulled back to eastern Tunisia. Its new positions were based on the old French fortifications of the Mareth Line, which were strengthened by German and Italian engineers so that the bulk of Rommel's forces could be freed for operations against the First Army's front.

Kasserine Pass

In the meantime, it was von Arnim who began the offensive on 14 February with a drive by 10th and 21st Panzer Divisions against the Faid Pass. This broke through the front-line held by American infantry, swiftly deploying to hull-down fire positions against a clumsy counterattack by 1st Armored's CCA, which was beaten off with the loss of 55 US AFVs. On the following day, a further ill-directed effort, this time by CCB, was defeated, during which 46 American tanks were captured or destroyed.

As von Arnim's spearheads moved westwards, Rommel struck north to link up with his attack at the Kasserine Pass, hoping to be able to exploit the breakthrough with a drive against the Allied rear and their key Algerian supply centres at Bone and Constantine. After punching through the Kasserine Pass on 18 February, Axis armour swept on to Thala and Sbiba where it was beaten off by defences with massive artillery support, backed up by the British 6th Armoured Division.

Despite having inflicted 10,000 casualties for the loss of only 2000, Rommel realized that it was time to break off the attack and by the end of the month, the Axis forces were back on their old front-lines.

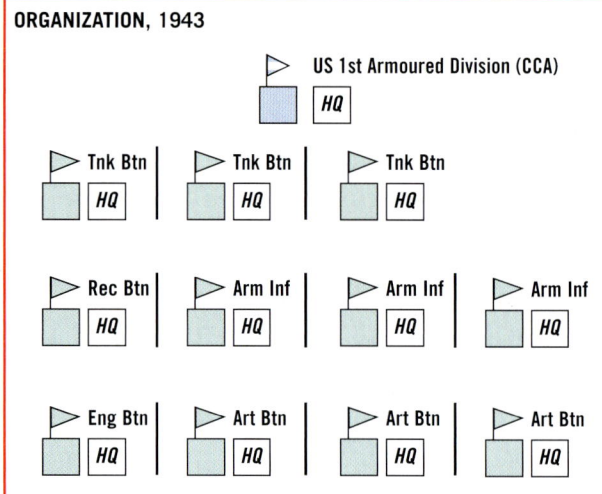

The DAK's final offensive was launched by three Panzer divisions with a total of 142 tanks and 200 guns against the Eighth Army's positions at Medinine. Good intelligence and reconnaissance ensured that Montgomery was well aware of what was coming and over 450 anti-tank guns (including the new 17pdrs) supported by 350 pieces of field and medium artillery broke up the attack, knocking out 50 of the Panzers.

Rommel was now desperately concerned by the overwhelming odds his men were facing and he handed over to von Arnim before returning to Germany to try to convince Hitler of just how critical the situation had become. Hitler refused to accept Rommel's bleak assessment of the chances of holding Tunisia and ordered him to take further sick leave.

▶ **M3A1 Light Tank**
First Army / 1st Armored Division / 12th Armored Regiment / 1st Battalion
In common with the Eighth Army, US forces in North Africa quickly appreciated the M3's qualities as a reconnaissance vehicle.

Specifications
Crew: 4
Weight: 12.9 tonnes (12.69 tons)
Length: 4.53m (14ft 10in)
Width: 2.24m (7ft 4in)
Height: 2.64m (8ft 8in)
Engine: 186.25kW (250hp) Continental W-670-9A 7-cylinder petrol
Speed: 58km/h (36mph)
Range: 110km (70 miles)
Armament: 1 x 37mm (1.5in) M6 gun, plus 3 x 7.62mm (0.3in) MGs (1 AA, 1 coaxial and 1 ball-mounted in hull front)
Radio: SCR508

NORTH AFRICA: 1940–43

▶ **Link-up**
M3 light tanks of US II Corps meet leading elements of the Eighth Army in Tunisia, April 1943. Axis forces in North Africa were now trapped in a shrinking pocket around Tunis, finally surrendering in mid-May.

Any chances that Rommel might be able to return to Africa were soon overtaken by the final Allied offensives, which squeezed the Axis forces into a shrinking perimeter around Tunis and Bizerta by mid-April. It was clearly only a matter of time before the enormous Allied superiority (1200 AFVs and 1500 guns to 130 tanks and 500 guns) would prove decisive. Despite some effective rearguard actions, notably by the handful of available Tigers, all Axis forces in North Africa surrendered on 12 May, ending almost three years of desert warfare.

▲ **M3 Medium Tank**

First Army / 1st Armored Division / 12th Armored Regiment / 2nd Battalion

If handled carefully, the M3 was still an effective medium tank in 1942–43, but US armoured units in North Africa initially lacked the necessary tactical expertise to overcome its design limitations.

Specifications

Crew: 7
Weight: 27.22 tonnes (26.7 tons)
Length: 5.64m (18ft 6in)
Width: 2.72m (8ft 11in)
Height: 3.12m (10ft 3in)
Engine: 253.5kW (340hp) Continental R-975-EC2 radial petrol
Speed: 42km/h (26mph)
Range: 193km (120 miles)
Armament: 1 x 75mm (2.9in) M2 or M3 gun, 1 x 37mm (1.5in) M4 or M5 gun, plus 4 x 7.62mm (0.3in) MGs (1 in commander's cupola, 1 coaxial and 2 fixed forward-firing)
Radio: SCR508

75

Chapter 4

Sicily and Italy: 1943–45

Winston Churchill was a keen proponent of the 'Mediterranean strategy' – exploiting the Axis collapse in North Africa by invading Italy, which he memorably (and inaccurately) termed 'the soft underbelly of Europe'. Allied hesitancy and German doggedness were to ensure that it was anything but soft. Indeed, to those who fought from Sicily to Turin it became 'the tough old gut'.

◀ **Desolation**
A Canadian Sherman stops in the ruins of Ortona, December 1943.

SICILY AND ITALY: 1943–45

Invasion of Sicily
10 July–17 August 1943

In early 1943, the Allies finally agreed that an invasion of France was not feasible that year and to adopt elements of Churchill's Mediterranean strategy, beginning with the invasion of Sicily.

THE ISLAND WAS DEFENDED by General Alfredo Guzzoni's Italian Sixth Army, with over 240,000 men, and the roughly 40,000 German troops of XIV Panzer Corps, comprising the *Hermann Göring* Panzer Division and 15th *Panzergrenadier* Division. The force had at least 270 AFVs and about 220 guns.

The invasion force, meanwhile, was built around General George Patton's US Seventh Army and Montgomery's Eighth Army, a total of 160,000 men, 600 tanks and 1800 guns carried by a fleet of 2500 vessels. Crucially, the fleet included battleships and cruisers, whose long-range gunfire support would do much to ensure the success of the landings on 10 July.

Airborne assault
Although the Allied airborne assault that was intended to block the routes to the beachheads was badly scattered, it did confuse the defenders, slowing their reaction to the landings. Patton's Seventh Army, including 2nd Armored Division, landed in the Gulf of Gela, in south-central Sicily, on a frontage of over 50km (31 miles). Montgomery's forces landed in south-eastern Sicily, also on a 50km (31-mile) front, with a 40km (25-mile) gap between the two armies. The plan called for the armies to link up and secure a large beachhead area before Eighth Army drove north to Messina, whilst Seventh Army covered its left flank and cleared the rest of the island.

Inevitably, there were setbacks and delays – whilst many Italian units were only too happy to surrender, others fought with surprising ferocity. As ever, German forces were able opponents, skilfully exploiting the terrain to delay the Allied advance and launching sharp counterattacks.

On 10 July, US forces around Gela fought off an Italian counterattack by the *Livorno* Division supported by tanks of the *Niscemi* Armoured Combat Group. Two days later the *Livorno* Division tried again in conjunction with the *Hermann Göring* Panzer Division, but this attack was also beaten off with the help of accurate naval gunfire support.

British advance
In contrast, British forces had only met light opposition in their capture of Syracuse, but on 13 July, they had to fight off counterattacks by infantry and ex-French R-35 tanks of General Giulio Porcinari's *Napoli* Infantry Division. (Elements of 4th Armoured Brigade captured Porcinari and his staff in this action.)

As the Allied advances continued, there was little tank-versus-tank action, but scores of small engagements such as those recorded in the report for 22 July of the US 2nd Armored Division (nicknamed 'Hell on Wheels'). 'Each defile was strongly defended by A-T weapons and machine-guns cleverly emplaced and protected by infantry. Each of these elements had to be reduced one by one. Not until surrounded by infantry and shelled by artillery and/or tanks was there any sign of surrender. For the most part, the

INDEPENDENT TANK BATTALIONS – SEVENTH ARMY

▷ US Seventh Army
HQ

▷ 70 Tk Btn   ▷ 753 Tk Btn   ▷ 756 Tk Btn
HQ   HQ   HQ

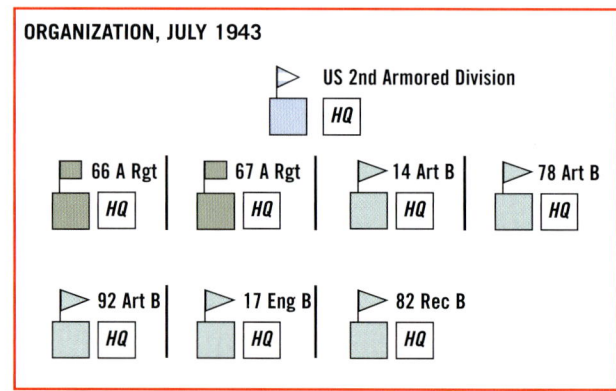

ORGANIZATION, JULY 1943

▷ US 2nd Armored Division
HQ

■ 66 A Rgt   ■ 67 A Rgt   ▷ 14 Art B   ▷ 78 Art B
HQ   HQ   HQ   HQ

▷ 92 Art B   ▷ 17 Eng B   ▷ 82 Rec B
HQ   HQ   HQ

SICILY AND ITALY: 1943–45

antitank guns were manned by Germans and the infantry furnished by the Italians.... Leading elements of the division on arrival at the pass 4 miles [6.5km] NE of San Guisseppe were held up by a determined defense in depth by A-T guns including German 88mm's [3.5in]. These guns were well emplaced in the sides of the canyons, cleverly concealed and in an extremely strong natural position. This resistance was overcome by flanking action of dismounted patrols, covered by artillery,

Specifications*
Crew: 5
Weight: 30.3 tonnes (29.82 tons)
Length: 5.89m (19ft 4in)
Width: 2.62m (8ft 7in)
Height: 2.74m (8ft 11in)
Engine: 298kW (400hp) Continental R975 C1 radial petrol
Speed: 34km/h (21mph)
Range: 193km (120 miles)
Armament: 1 x 75mm (2.9in) M3 gun, plus 2 x 7.62mm (0.3in) MGs (1 coaxial, 1 ball-mounted in hull front)
Radio: SCR 508

*Data is for standard M4 - the dozer attachment increased weight, length and width, but no figures are available

▲ M4 Dozer
US Seventh Army / 2nd Armored Division / 17th Armored Engineer Battalion
The Sherman's great virtue was its adaptability, an early modification being the installation of a dozer blade. The resulting vehicle proved invaluable for a host of tasks, from clearing rubble-strewn roads to acting as a recovery vehicle.

▲ M4A1 Sherman Medium Tank
US Seventh Army / 2nd Armored Division / 66th Armored Regiment
Successful operations in Sicily demonstrated the vast improvement in US armoured formations since the debacle at Kasserine Pass in North Africa only a few months earlier.

Specifications
Crew: 5
Weight: 30.3 tonnes (29.82 tons)
Length: 5.84m (19ft 2in)
Width: 2.62m (8ft 7in)
Height: 2.74m (8ft 11in)
Engine: 298kW (400hp) Continental R975 C1 radial petrol
Speed: 34km/h (21mph)
Range: 193km (120 miles)
Armament: 1 x 75mm (2.9in) M3 gun, plus 2 x 7.62mm (0.3in) MGs (1 coaxial, 1 ball-mounted in hull front)
Radio: SCR 508

SICILY AND ITALY: 1943–45

▲ **Safely ashore**
An M4A1 Sherman of 2nd Armored Division departs from Red Beach 2, Sicily.

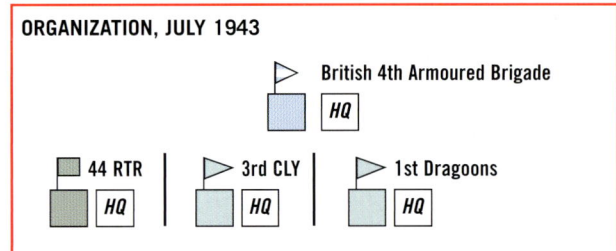

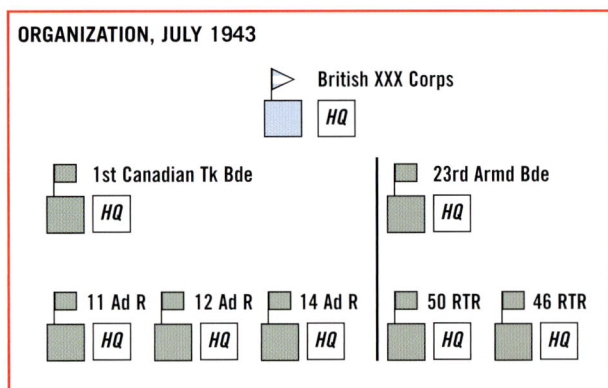

tank and supporting cannon gun fire. In the meantime, reconnaissance was being pushed around the flanks to determine routes to by-pass this defile.'

After much bitter fighting, Sicily was finally cleared of Axis forces on 17 August 1943. In a brilliantly executed operation, the Germans evacuated 40,000 men, hundreds of vehicles, including 44 tanks, together with thousands of tonnes of ammunition and supplies under constant air attack. The Sicilian campaign had been a sharp reminder of the importance of good logistical support in successful armoured actions. The US 2nd Armored Division's report ruefully concluded: 'The operation against PALERMO served to emphasize the tremendous supply problem involved in sustaining an armored division on the move and in action. It is estimated

▲ **105mm Howitzer Motor Carriage (HMC) M7 (Priest)**
US Seventh Army / 2nd Armored Division / 14th Armored Field Artillery Battalion

This was a heavily modified M3 medium tank hull, fitted with the standard US 105mm (4.1in) howitzer in a thinly armoured, open-topped fighting compartment. In British service, it was referred to as the Priest due to the pulpit-like machine-gun position.

Specifications
Crew: 7
Weight: 26.01 tonnes (25.6 tons)
Length: 6.02m (19ft 9in)
Width: 2.88m (9ft 5in)
Height: 2.54m (8ft 4in)
Engine: 298kW (400hp) Continental R975 C1 radial petrol
Speed: 41.8km/h (26mph)
Range: 201km (125 miles)
Armament: 1 x 105mm (4.1in) M1A2 howitzer, plus 1 x 12.7mm (0.5in) HMG on 'pulpit' AA mount.
Radio: SCR 608

SICILY AND ITALY: 1943–45

that the organic vehicles within an armored division can keep the division supplied as long as the Army rail or truck head is within 30 miles [48km] of the combat elements and a reasonable road net exists.

In 1942, red-and-white recognition stripes were applied to British AFVs. These were retained until at least mid-1944, when they were largely superseded by the Allied white star.

▲ Bishop SP 25pdr
Eighth Army / XIII Corps / 24th Armoured Field Regiment RA

This was an early attempt to provide a fully tracked SP version of the 25pdr gun/howitzer, based on the Valentine's hull. It is possible that the original intention was to use the vehicle as a stop-gap tank destroyer until the 6pdr-armed Crusaders and Valentines were ready. This would account for the awkward gun mounting in a fixed, fully enclosed armoured fighting compartment with limited traverse. More importantly, the mounting severely restricted elevation, which almost halved the gun's maximum range – 5850m (6400 yards), compared with the towed gun's 10,900m (11,925 yards). The 100 or so Bishops produced entered service in the first half of 1942 and were used throughout the remainder of the desert war and in Sicily, before being replaced by US-supplied Priest SP 105mm (4.1in) howitzers during the early stages of the Italian campaign.

Specifications
Crew: 4
Weight: 20.32 tonnes (20 tons)
Length: 5.62m (18ft 6in)
Width: 2.77m (9ft 1in)
Height: 3.05m (10ft)
Engine: 98kW (131hp) AEC 6-cylinder diesel
Speed: 24km/h (15mph)
Range: 177km (110 miles)
Armament: 1 x 87.6mm (3.45in) 25pdr gun/howitzer
Radio: Wireless Set No. 19

▲ T19 105mm Howitzer Motor Carriage (HMC)
US Seventh Army / 2nd Armored Division / 92nd Armored Field Artillery Battalion

The T19 HMC was an M3 halftrack mounting the M2A1 105mm (4.1in) howitzer. The M3 could only stow eight rounds of 105mm (4.1in) ammunition, whilst the chassis was only just able to withstand the howitzer's weight and recoil. The type was phased out as more Priests became available.

Specifications
Crew: 6
Weight: 9.1 tonnes (8.95 tons)
Length: 6.16 m (20ft 2.5in)
Width: 1.96m (6ft 5in)
Height: 2.3m (7ft 6.5in)
Engine: 109.5kW (147hp) White 160AX 6-cylinder petrol
Speed: 72km/h (45mph)
Range: 320km (200 miles)
Armament: 1 x 105mm (4.1in) M1A2 howitzer
Radio: SCR 608

SICILY AND ITALY: 1943–45

▲ M4 81mm Motor Mortar Carriage (MMC)
US Seventh Army / 2nd Armored Division / 41st Armored Infantry Regiment

The M4 was a version of the M2 halftrack fitted with the M1 81mm (3.19in) mortar, which was intended to be fired dismounted but could also be fired to the rear from a mounting inside the vehicle.

Specifications
Crew: 6
Weight: 7.97 tonnes (7.84 tons)
Length: 6.01m (19ft 8.6in)
Width: 1.96m (6ft 5in)
Height: 2.27m (7ft 5in)
Engine: 109.5kW (147hp) White 160AX 6-cylinder petrol
Speed: 72km/h (45mph)
Range: 320km (200 miles)
Armament: 1 x 81mm (3.19in) M1 mortar, plus 1 x 12.7mm (0.5in) HMG
Radio: n/k

▲ Car, Heavy Utility, 4x2, Ford C11 ADF
Eighth Army / HQ XXX Corps

The C11 was a slightly modified version of the civilian Ford Fordor station wagon and was widely used as a staff car. This example flies a lieutenant-general's pennant, indicating its use by the corps commander, Sir Oliver Leese.

Specifications
Crew: 1
Weight: 1.8 tonnes (1.78 tons)
Length: 4.83m (16ft 2in)
Width: 2.01m (6ft 7in)
Height: 1.83m (6ft)
Engine: 70.8kW (95hp) Ford Mercury V8 petrol
Speed: 89km/h (55mph) (estimated)
Range: 370km (230 miles) (estimated)

▶ Car, Heavy Utility, 4x4 (FWD) Humber
HQ Eighth Army

Generally referred to as the 'Humber Box', this was the only British-built 4x4 vehicle of its type to enter British Army service during the war years, seeing extensive service as a staff car.

Specifications
Crew: 1
Weight: 2.41 tonnes (2.37 tons)
Length: 4.29m (14ft 1in)
Width: 1.88m (6ft 2in)
Height: 1.96m (6ft 5in)
Engine: 63.4kW (85hp) Humber 6-cylinder petrol
Speed: 80km/h (50mph) (estimated)
Range: 300km (186 miles) (estimated)

SICILY AND ITALY: 1943–45

As this division landed with a very limited number of trucks due to shortage of shipping, it was able to maintain itself only by a close margin. All trucks hauled 24 hours a day, being forced to draw from beach dumps. Due to the rapid movement of the division, the distance from these dumps increased until it reached 140 miles [224km].

Fortunately, ammunition requirements for the operation were not heavy. Had the action been sustained and the demand for ammunition tonnage been heavy, it would have been impossible to have supplied the division with both gasoline and ammunition with the trucks available. The entire operation would have been seriously impeded and might have been entirely jeopardized.'

Mountain warning

The Sicilian campaign also gave a taste of some of the conditions that were to become all too familiar in Italy itself. The Axis defences in the mountainous north-east of the island (the Etna Line) had largely relegated Allied armour to acting in the infantry support role or as improvised artillery. It was the shape of things to come.

Specifications
Crew: 1
Weight: n/k
Length: 3.98m (13ft 1in)
Width: 1.6m (5ft 3in)
Height: 1.96m (6ft 5in)
Engine: 21.97kW (29.5hp) Austin 4-cylinder petrol
Speed: 80km/h (50mph) (estimated)
Range: 300km (186 miles) (estimated)

▲ **Austin 10 Light Utility Truck**
Eighth Army / XIII Corps / HQ 50th (Northumbrian) Division
The Austin 10 utility truck (better known as the Austin Tilly) was the military version of the Austin 10 saloon and was widely used for a host of light transport duties. About 30,000 vehicles were built.

Specifications
Crew: 4 or 5
Weight: 2.17 tonnes (2.14 tons)
Length: 4.29m (14ft 1in)
Width: 1.88m (6ft 2in)
Height: 1.89m (6ft 2.4in)
Engine: 64kW (86hp) Humber 6-cylinder petrol
Speed: 75km/h (46mph)
Range: 500km (310 miles)

▲ **Humber Snipe Light Utility Truck**
Eighth Army / XXX Corps / HQ 51st (Highland) Division
Humber built large numbers of light utility trucks based on the civilian Snipe. This is the FFW (Fitted For Wireless) version, with a detachable body for use as a ground wireless station.

SICILY AND ITALY: 1943–45

▼ US 2nd Armored Division, Combat Command A – Sicily, July 1943

The US armoured divisions were especially flexible formations. The abolition of their armoured brigades in March 1942 and their replacement by the combat command structure allowed units to be switched between combat commands to meet the changing needs of each particular mission.

Combat Command HQ (10 x M3 half-tracks, 3 x M3A1 Stuarts)

66th Armored Rgt (detachment)

Battalion HQ (2 x M3 half-tracks, 2 x M4 Shermans)

Tank Company x 3

Company HQ (2 x M4 Shermans, 1 x M3 half-track, 1 x recovery vehicle)

1 Tank Platoon (5 x M4 Shermans)

2 Tank Platoon (5 x M4 Shermans)

3 Tank Platoon (5 x M4 Shermans)

Field Artillery Battery, from 78th Armored Field Artillery Battalion (3 x M3 half-tracks, 6 x HMC M7s)

Battalion HQ **Fire Control Section**

SICILY AND ITALY: 1943–45

▲ **DUKW**

US Seventh Army / 2nd Armored Division / Quartermaster Corps (QMC) Battalion

The DUKW was an amphibious version of the General Motors 2½-ton cargo truck, which entered service in 1942. It proved to be a highly successful means of delivering supplies directly to shore from ships at sea.

Specifications

Crew: 1
Weight: 6.75 tonnes (6.64 tons)
Length: 9.75m (32ft)
Width: 2.51m (8ft 2.9in)
Height: 2.69m (8ft 10in)
Engine: 68.2kW (91.5hp) GMC Model 270 petrol
Speed: 80km/h (50mph)
Range: n/k

Into Italy: Salerno landings
SEPTEMBER–DECEMBER 1943

The realization that there was no prospect of defeating the Allied forces in Sicily was the final blow to Mussolini's already shaky regime. He was deposed and arrested on 25 July whilst Marshal Pietro Badoglio's new Italian government began negotiations with the Allies with a view to changing sides.

UNFORTUNATELY, DIPLOMATIC leaks alerted the Germans, who massively reinforced their units in Italy and disarmed most Italian forces as soon as the Badoglio government signed an armistice on 3 September. Soon after the armistice, Mussolini was rescued from his imprisonment in the Campo Imperatore Hotel at Gran Sasso by German Special Forces and installed as head of state of the puppet Italian Social Republic in northern Italy.

Landings

On the same day that the armistice was signed in Cassibile, Sicily, Montgomery's forces made unopposed landings at Reggio di Calabria, whilst General Mark Clark's US Fifth Army (including British units) landed at Salerno, south of Naples, on 9 September. The plan was daring but flawed – Fifth Army's landings were made on a very broad 56km (35-mile) front, using only three assault divisions, and the army's two corps were widely separated. This was potentially a recipe for disaster, especially as 16th Panzer Division was deployed within easy reach of the beachheads and Clark had specifically ruled out any preliminary naval bombardment.

By 12 September, the Germans were able to launch a concentrated assault by two *Panzergrenadier* and four Panzer divisions which destroyed a US infantry battalion and came close to overrunning the entire beachhead before it was halted by massed artillery and air attacks, backed up by naval gunfire support from the battleships *Warspite* and *Valiant*. As soon as it became clear that there was no realistic chance of driving the Allies back into the sea, the Germans skilfully disengaged and withdrew to the north.

With the Salerno beachhead secure, Fifth Army began its attack towards Naples on 19 September. The Eighth Army's 7th Armoured Division was ordered to take Naples, which fell to A Squadron, King's Dragoon Guards, on 1 October after the occupying German forces withdrew following a

SICILY AND ITALY: 1943–45

popular uprising in the city. By 6 October, Fifth Army had closed up to the line of the River Volturno, the first of many German defence lines that were to slow the Allied advance up the Italian peninsula to a long, bloody crawl. Narrow Italian roads and bridges were obvious targets for demolitions and it was said with feeling that the best Allied weapon of the campaign was the bulldozer rather than the tank.

On 6 November, Hitler appointed the *Luftwaffe*'s Field-Marshal Albert Kesselring as commander-in-chief in Italy, approving his plans for a 'forward defence' of the country. The Volturno Line was defended just long enough to give time for the completion of the Winter Line, the main fortified zone south of Rome.

It was not until mid-January 1944 that the Allies managed to fight their way through to the key element of the Winter Line defences, the Gustav Line. This made use of some of the most difficult terrain in Europe, strengthened by an array of gun pits, concrete bunkers, machine-gun turrets, barbed wire and minefields. It was ably defended by 15 German divisions until late May 1944 and was the setting for the four battles of Monte Cassino.

▲ Ford GPA Seagoing Jeep (Seep)
US Seventh Army / 540th Engineer Shore Regiment

The GPA was intended to be a fully amphibious version of the jeep, but its greater weight and poor waterborne performance led to its withdrawal from front-line US service shortly after the campaign in Sicily.

Specifications

Crew: 1	Height: 1.75m (5ft 9in)
Weight: 1.63 tonnes (1.6 tons)	Engine: 44.7kW (60hp) Ford 4-cylinder petrol
Length: 4.62m (15ft 2in)	Speed: 105km/h (65mph)
Width: 1.63m (5ft 4in)	Range: n/k

▲ M2A1 Halftrack
US Fifth Army / VI Corps / 1st Armored Division / 11th Armored Infantry Battalion

The M2A1 was fitted with the armoured M49 machine-gun ring mount over the right-hand front seat. Two or three fixed pintle mounts for 7.62mm (0.3in) machine guns were often fitted at the unit level in the field.

Specifications

Crew: 2 plus 8 passengers	6-cylinder petrol
Weight: 8.89 tonnes (8.75 tons)	Speed: 72km/h (45mph)
Length: 6.14m (20ft 1in)	Range: 320km (200 miles)
Width: 2.22m (7ft 3in)	Armament: 1 x 12.7mm (0.5in) HMG,
Height: 2.69m (8ft 10in)	plus 2 x 7.62mm (0.3in) MGs
Engine: 109.5kW (147hp) White 160AX	

SICILY AND ITALY: 1943–45

▲ **Amphibious landings**
American Sherman tanks splash through the surf at Salerno. Although the Sherman was inferior to German tanks in terms of firepower and armour, it was mechanically reliable and produced in large numbers.

▲ **M8 Armoured Car**
US Fifth Army / IV Corps / 1st Armored Division / Cavalry Reconnaissance Squadron
Many users felt that the M8's armour was inadequate and that the vehicle was especially vulnerable to mines. Crews often lined the floor of the fighting compartment with sandbags to give a measure of protection against mines.

▶ **Willys MB Jeep**
US Seventh Army / 45th Infantry Division / 45th Reconnaissance Troop
The versatile jeep appeared in many guises during (and long after) the war. This reconnaissance vehicle is armed with a pintle-mounted 12.7mm (0.5in) heavy machine gun.

Specifications
Crew: 4
Weight: 8.12 tonnes (8 tons)
Length: 5m (16ft 5in)
Width: 2.54m (8ft 4in)
Height: 2.25m (7ft 5in)
Engine: 82kW (110hp) Hercules JXD 6-cylinder petrol
Speed: 89km/h (55mph)
Range: 563km (350 miles)
Armament: 1 x 37mm (1.5in) M6 gun, plus 1 x 12.7mm (0.5in) AA HMG and 1 x coaxial 7.62mm (0.3in) MG
Radio: SCR508

Specifications
Crew: 1 plus 2/3 passengers
Weight: 1.04 tonnes (1.02 tons)
Length: 3.33m (10ft 11in)
Width: 1.57m (5ft 2in)
Height: 1.83m (6ft)
Engine: 40.23kW (54hp) L head 134 I4 4-cylinder petrol
Speed: 105km/h (65mph)
Range: 482.8km (300 miles)
Armament: 1 x 12.7mm (0.5in) HMG or 1 x 7.62mm (0.3in) MG

SICILY AND ITALY: 1943–45

Anzio
JANUARY–MAY 1944

As Allied forces struggled to punch through the Winter Line, a further effort was made to break the stalemate, by means of an amphibious assault at Anzio, south of Rome.

A SHORTAGE OF TANK LANDING SHIPS, which were urgently required for the preparations for the Normandy landings, limited the scale of the initial assault to little more than two divisions. This force, comprising the British 1st Infantry Division and the US 3rd Infantry Division, landed practically unopposed on 22 January 1944 but passively awaited reinforcements instead of advancing inland. As usual, Kesselring reacted swiftly and within three days the beachheads were sealed off by 40,000 German troops, and the chance of a quick breakthrough had been lost. Yet the terrain was once again a limiting

▲ **M4A3 Sherman Medium Tank**
US Fifth Army / VI Corps / 1st Armored Division

As the Italian campaign dragged on, increasingly frequent encounters with Panthers, Tigers and powerful tank destroyers convinced many Sherman crews that their vehicles were becoming obsolete. US and British officials were reluctant to accept this and much time was lost before work got under way on a successor.

▶ **Staghound Mark I Armoured Car**
Eighth Army / 2nd New Zealand Division / Divisional Cavalry Regiment

'Staghound' was the British designation for the Ford T17E1 armoured car. The vehicle was not adopted by the US Army and only saw service with British and Commonwealth forces.

Specifications
Crew: 5
Weight: 30.3 tonnes (29.82 tons)
Length: 5.9m (19ft 4in)
Width: 2.62m (8ft 7in)
Height: 2.74m (8ft 11in)
Engine: 372.5kW (500hp) Ford GAA 8-cylinder petrol
Speed: 42km/h (26mph)
Range: 210km (130 miles)
Armament: 1 x 75mm (2.9in) M3 gun, plus 1 x 12.7mm (0.5in) AA HMG and 2 x 7.62mm (0.3in) MGs (1 coaxial, 1 ball-mounted in hull front)
Radio: SCR 508

Specifications
Crew: 5
Weight: 14.12 tonnes (13.9 tons)
Length: 5.49m (18ft)
Width: 2.69m (8ft 10in)
Height: 2.36m (7ft 9in)
Engine: 2 x 72kW (97hp) GMC 270 CID 6-cylinder in-line petrol
Speed: 89km/h (55mph)
Range: 724km (450 miles)
Armament: 1 x 37mm (1.5in) M6 gun plus 3 x 7.62mm (0.3in) MGs (1 AA, 1 coaxial and 1 ball-mounted in hull front)
Radio: Wireless Set No. 19

SICILY AND ITALY: 1943–45

factor – in the initial stages of the landings a single Sherman of the Scots Greys commanded by Sergeant McMeeking was the only tank able to get through the marshy ground at the beach exits. He arrived at the forward infantry positions just in time to break up a German counterattack by single-handedly destroying four Panzer IVs in quick succession.

Stalemate

Both sides reinforced the area, and a series of attacks and counterattacks in February ended in another bloody stalemate that lasted until late May when the Allies finally broke through the Gustav Line in the Fourth Battle of Monte Cassino. At this point, on 23 May, the US VI Corps, comprising three US and two British divisions, broke out of the Anzio beachhead. The opposing German Fourteenth Army had been weakened by the need to reinforce the Gustav Line but still managed to inflict heavy losses – on the first day of the offensive 1st Armored Division lost 100 tanks whilst 3rd Infantry Division suffered 955 casualties, the highest losses sustained in a single day by any US division in World War II.

By 25 May, the seven divisions of the German Tenth Army were in full retreat from the Gustav Line and VI Corps was poised to cut off their retreat towards Rome along Route 6. At this point, Clark

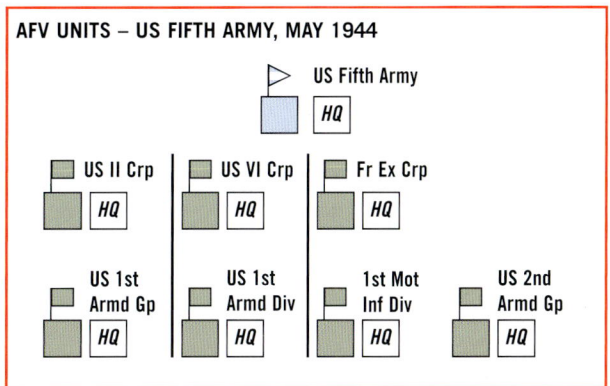

▶ **Mud and guts**
Shermans and British infantry move up through the mud to the front-line, 1944.

▲ **M2 Halftrack**
US Fifth Army / VI Corps / 1st Armored Division
This M2 shows some of the stowage arrangements – a side compartment, in this case packed with ammunition boxes, and an external rack for anti-tank mines.

Specifications
Crew: 2 plus 8 passengers
Weight: 8.7 tonnes (8.56 tons)
Length: 5.96m (19ft 6in)
Width: 1.96m (6ft 5in)
Height: 2.3m (7ft 6in)
Engine: 109.5kW (147hp) White 160AX
  6-cylinder petrol
Speed: 72km/h (45mph)
Range: 320km (200 miles)
Armament: 1 x 12.7mm (0.5in) HMG

SICILY AND ITALY: 1943–45

ordered General Lucian Truscott, commanding VI Corps, to abandon the advance on Route 6 and move directly on Rome. It seems likely that this was simply down to Clark's personal ambition to go down in history as the liberator of the Eternal City. Whatever the motivation, it is certain that a great opportunity was missed, and General Heinrich von Vietinghoff's Tenth Army was able to retreat to the next line of defence, the Trasimene Line, where it linked up with Fourteenth Army, under General Joachim Lemelsen, before making a fighting withdrawal to the Gothic Line north of Florence.

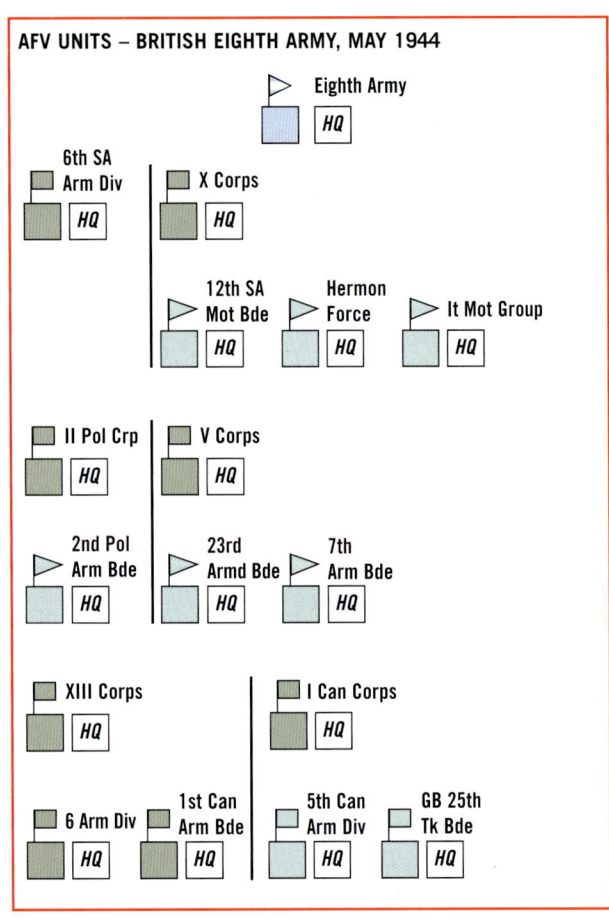

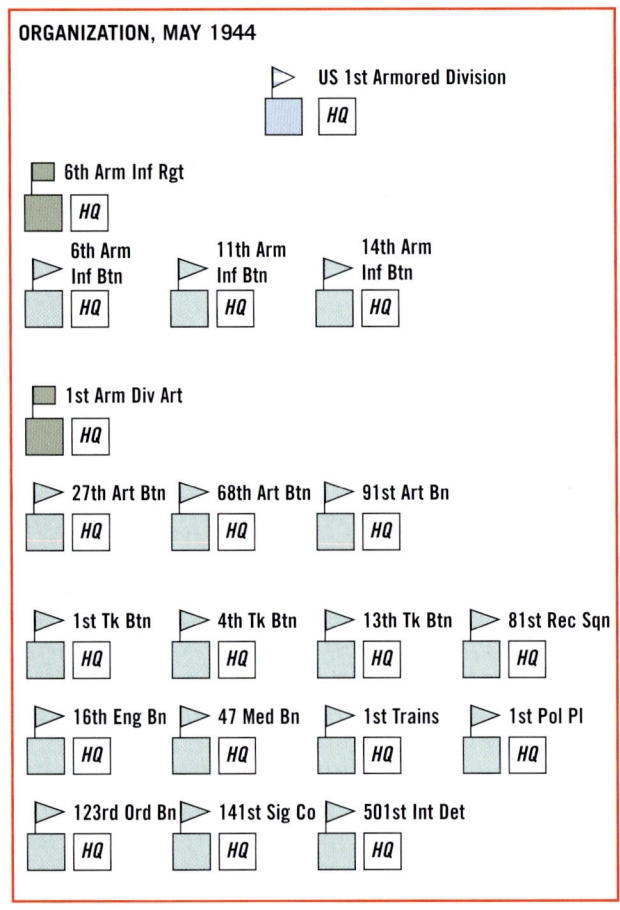

▶ **Staghound AA Armoured Car**

Eighth Army / 2nd New Zealand Division / Divisional Cavalry Regiment

This was the Staghound Mark I fitted with an open-topped Frazer-Nash turret mounting two 12.7mm (0.5in) M2 Browning heavy machine guns with 1305 rounds per gun. Allied air superiority meant that the type was most commonly used in the fire support role. (The high elevation of the guns often proved useful in the mountainous terrain encountered in much of Italy.)

Specifications

Crew: 5
Weight: 12.05 tonnes (11.86 tons)
Length: 5.43m (17ft 10in)
Width: 2.69m (8ft 10in)
Height: 2.42m (7ft 11in)
Engine: 2 x 72kW (97hp) GMC 270 CID 6-cylinder in-line petrol
Speed: 89km/h (55mph)
Range: 320km (200 miles)
Armament: 2 x 12.7mm (0.5in) HMGs
Radio: Wireless Set No. 19

SICILY AND ITALY: 1943–45

Gothic Line
SEPTEMBER 1944

Rome fell to units of Clark's US Fifth Army on 4 June, only two days before the Italian campaign was overshadowed by the Normandy landings.

THE GOTHIC LINE was a belt of fortifications 16km (10 miles) deep, running from south of La Spezia on the Mediterranean coast to the Adriatic near Ravenna. It passed through the superb defensive terrain of the Apennine mountains, which ran unbroken nearly from coast to coast, 80km (50 miles) deep, with peaks rising to 2100m (6890ft).

The line's defences included numerous concrete-reinforced gun pits and trenches, 2376 machine-gun nests with interlocking fields of fire, 479 anti-tank gun, mortar and assault-gun positions, 120km (75 miles) of barbed wire and anti-tank ditches. Some positions included the imposing *Pantherturm* – a Panther tank turret, complete with 75mm (2.9in) gun, built on to a concrete or armoured emplacement to create a deadly anti-tank position.

Attacks on the outposts of the Gothic Line began on 25 August, but it was not until 12 September that Allied forces were able to close up on the line itself. The experience of Allied armour in Italy was typified by the ordeal of 21st Tank Brigade, which was heavily involved in innumerable infantry support actions throughout September. These imposed immense stress on men and tanks alike: a tank driver spoke of

▶ **Daimler Scout Car Mark IA (Dingo)**
US Fifth Army / XIII Corps / 6th Armoured Division / 1st Derbyshire Yeomanry
The Daimler Scout Car, almost invariably known as the Dingo, was one of the most successful light AFVs of the war. It was an ideal reconnaissance vehicle with a quiet engine, high speed in forward and reverse gears and a low silhouette.

Specifications
Crew: 2
Weight: 3.22 tonnes (3.2 tons)
Length: 3.23m (10ft 5in)
Width: 1.72m (5ft 8in)
Height: 1.5m (4ft 11in)
Engine: 41kW (55hp) Daimler 6-cylinder petrol
Speed: 89km/h (55mph)
Range: 322km (200 miles)
Armament: 1 x 7.7mm (0.303in) Bren MG
Radio: Wireless Set No. 19

▶ **Daimler Armoured Car**
US Fifth Army / XIII Corps / 6th Armoured Division / 1st Derbyshire Yeomanry
The Daimler Armoured Car was essentially an enlarged development of the Daimler Scout Car, fitted with the two-man turret of the Tetrarch light tank. Almost 2700 vehicles were produced and many remained in service well after the end of the war.

Specifications
Crew: 3
Weight: 7.62 tonnes (7.5 tons)
Length: 3.96m (13ft)
Width: 2.44m (8ft)
Height: 2.24m (7ft 4in)
Engine: 71kW (95hp) Daimler 6-cylinder petrol
Speed: 80km/h (50mph)
Range: 330km (205 miles)
Armament: 1 x 40mm (1.57in) 2pdr OQF gun, plus 1 x coaxial 7.92mm (0.31in) Besa MG
Radio: Wireless Set No. 19

SICILY AND ITALY: 1943–45

▲ **Look out for Panzerfausts!**
A Sherman of the Canadian Three Rivers Regiment advances through the ruins of Ortona.

'Hideous nights pressed on days of horror – we lost men, we lost tanks, we almost lost hope of survival.'

The brigade, part of I Canadian Corps within the Eighth Army, lost 52 Churchills, 29 Shermans and four Stuarts from the reconnaissance squadron. During the month it fired off 2828 armour-piercing rounds, 9632 HE shells and just under one million rounds of machine-gun ammunition.

Weather and terrain

Weather conditions and the terrain posed as many problems as the enemy – cloud cover often ruled out close air support, whilst the steep mountain slopes confined Allied armour to the muddy valley roads. These were heavily mined and booby-trapped besides being covered by German artillery and anti-tank guns. Even Allied artillery was not much use, the poor roads, washed-out bridges and mud limited the numbers of shells that could be brought forward, and most German bunkers were proof against all but a direct hit. Allied forces had to resort to infantry attacks under cover of artillery bombardments to take the German positions one at a time with grenades and small-arms fire, a very slow way indeed to advance. A locally produced German forces' news-sheet described these attacks:

'The Americans use quasi-Indian tactics. They search for the boundary between battalions or regiments, they look for gaps between our strongest points, they look for the steepest mountain passages (guided by treacherous civilians). They infiltrate through these passages with a patrol, a platoon at first, mostly at dusk. At night they reinforce the infiltrated units, and in the morning they are often in the rear of a German unit, which is simultaneously being attacked from the flanks …'

SICILY AND ITALY: 1943–45

▼ Armoured Car Squadron, 1st Derbyshire Yeomanry, 1944

The armoured car squadron was a highly mobile reconnaissance force. Whilst always attempting to carry out its missions by stealth, it had sufficient strength to fight its way through when necessary.

Squadron HQ (4 x Daimler AC, 1 x Daimler scout car)

1 Troop (2 x Daimler AC, 2 x Daimler scout car)

2 Troop (2 x Daimler AC, 2 x Daimler scout car)

3 Troop (2 x Daimler AC, 2 x Daimler scout car)

4 Troop (2 x Daimler AC, 2 x Daimler scout car)

5 Troop (2 x Daimler AC, 2 x Daimler scout car)

Heavy Troop (1 x Daimler scout car, 2 x AEC AC)

Support Troop (1 x Daimler scout car, 3 x M2/M3 half-tracks)

AA Troop (5 x Staghound AC)

SICILY AND ITALY: 1943–45

Battering down the door

It took a further week's hard fighting to break through the main defences, only for the advance to bog down. Torrential rain caused mudslides that frequently blocked roads and tracks, creating a logistical nightmare. Although the Allies were through the mountains, the Lombardy plains were waterlogged and Eighth Army found itself confronted, as in the previous autumn, by a succession of swollen rivers running across its line of advance. Once again, conditions (and losses totalling 480 AFVs) prevented Eighth Army's armour from exploiting the breakthrough. Both sides settled into their positions for what was to be the war's final winter.

▲ M5 High-Speed Tractor
US Fifth Army / II Corps / 88th Infantry Division / 337th Field Artillery Battalion

The M5 entered service with the US Army in 1942 and was extensively used by field artillery battalions.

Specifications

Crew: 1
Weight: 13.8 tonnes (13.58 tons)
Length: 5.03m (16ft 6in)
Width: 2.54m (8ft 4in)
Height: 2.69m (8ft 10in)
Engine: 154kW (207hp) Continental R6572 6-cylinder petrol
Speed: 48km/h (30 mph)
Range: 290km (180 miles)

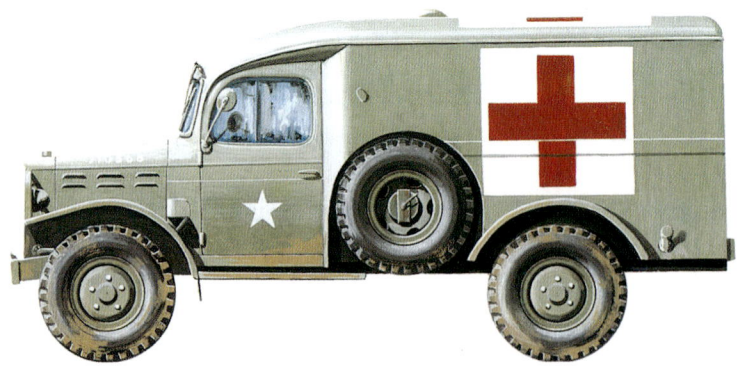

▲ WC54 Dodge 4x4 Ambulance
US Fifth Army / IV Corps / 1st Armored Division / Armored Medical Battalion

A total of 26,000 of these ambulances were produced between 1942 and 1944. Each could carry four stretchers or seven seated casualties.

Specifications

Crew: 2
Weight: 2.04 tonnes (2.01 tons)
Length: 4.67m (15ft 4in)
Width: 1.93m (6ft 4in)
Height: 2.13m (7ft 10in)
Engine: 68.6kW (92hp) Dodge T2155 6-cylinder petrol
Speed: 89km/h (55mph) (estimated)
Range: 402km (250 miles) (estimated)

SICILY AND ITALY: 1943–45

▲ **Shelling the enemy**
An M36 tank destroyer shells German positions on the plains of Lombardy, autumn 1944.

▲ **3in Gun Motor Carriage (GMC) M10**
US Fifth Army / VI Corps / 3rd Infantry Division / 601st Tank Destroyer Battalion

The M10 was the first US fully tracked tank destroyer design to enter service, utilizing the chassis of the Sherman with a new, thinly armoured superstructure and an open-topped turret mounting a 76mm (3in) M7 high-velocity gun.

Specifications

Crew: 5
Weight: 29.05 tonnes (28.6 tons)
Length: 5.82m (19ft 1in)
Width: 3.05m (10ft)
Height: 2.49m (8ft 2in)
Engine: 305.45kW (410hp) General Motors 6046 12-cylinder twin in-line diesel
Speed: 48km/h (30mph)
Range: 320km (200 miles)
Armament: 1 x 76mm (3in) M7 gun, plus 1 x 12.7mm (0.5in) HMG on AA mount
Radio: SCR610

SICILY AND ITALY: 1943–45

▶ Universal Carrier
US Fifth Army / 6th Armoured Division / 61st Infantry Brigade / 2nd Battalion Rifle Brigade

The Universal Carrier was found to be as useful in Italy as in every other theatre of war. The 13-vehicle Carrier Platoons in each infantry battalion were always in demand for a bewildering variety of tasks.

Specifications	
Crew: 3	Engine: 63.4kW (85hp) Ford V8
Weight: 4.06 tonnes (4 tons)	8-cylinder petrol
Length: 3.76m (12ft 4in)	Speed: 52km/h (32mph)
Width: 2.11m (6ft 11in)	Range: 258km (160 miles)
Height: 1.63m (5ft 4in)	Armament: 2 x 7.7mm (0.303in) Bren MGs

▶ Dodge WC58 Command Reconnaissance Radio Car
US Fifth Army / IV Corps / 1st Armored Division

This dedicated radio vehicle was a relative rarity – just over 2300 were built between 1942 and 1945.

Specifications	
Crew: 3	Engine: 68.54kW (92hp) Dodge T214
Weight: 2.42 tonnes (2.38 tons)	6-cylinder petrol
Length: 4.46m (14ft 7in)	Speed: 89km/h (55mph)
Width: 2m (6ft 7in)	Range: 384km (240 miles)
Height: 2.07m (6ft 9in)	Radio: SCR506-AFII

▲ Dodge WC51 Weapons Carrier
US Fifth Army / IV Corps / 1st Armored Division

Although officially designated as weapons carriers, the 98,000 or so WC51s produced during the war were mainly used for general transport duties.

Specifications	
Crew: 1	Engine: 68.54kW (92hp) Dodge T214
Weight: 3.3 tonnes (3.24 tons)	6-cylinder petrol
Length: 4.47m (14ft 8in)	Speed: 89km/h (55mph)
Width: 2.1m (6ft 10in)	Range: 384km (240 miles)
Height: 2.15m (7ft)	

SICILY AND ITALY: 1943–45

▲ **Sherman Firefly**
Eighth Army / Polish II Corps / 2nd Armoured Brigade / 1st Krechowiecki Lancers / 2nd Squadron
Roughly 100 Fireflies were sent to Italy, where they were issued to the 2nd, 4th and 7th New Zealand and 5th Canadian armoured brigades, as well as the Polish 2nd Armoured Brigade.

Specifications
Crew: 4
Weight: 32.7 tonnes (32.18 tons)
Length: 7.85m (25ft 9in)
Width: 2.67m (8ft 9in)
Height: 2.74m (8ft 11in)
Engine: 316.6kW (425hp) Chrysler Multibank A57 petrol
Speed: 40km/h (25mph)
Range: 161km (100 miles)
Armament: 1 x 76mm (3in) 17pdr OQF, plus 1 coaxial 7.62mm (0.3in) MG
Radio: Wireless Set No. 19

The last battles
MARCH–MAY 1945

By late March 1945, Allied strength amounted to 17 divisions plus eight independent brigades – equivalent overall to almost 20 divisions. These forces included four groups of volunteers from the former Italian Army, equipped and trained by the British.

THEY FACED 21 MUCH WEAKER German divisions and four Italian Fascist divisions. Three of the Italian divisions formed part of the Ligurian Army under Marshal Graziani, covering La Spezia and Genoa, whilst the fourth came under command of Fourteenth Army and was deployed in a sector thought less likely to be attacked.

Final offensive
The Allied offensive began on 9 April 1945, preceded by carpet-bombing (using 175,000 fragmentation bombs) and heavy artillery bombardment. Wasps – flamethrower-armed Universal Carriers – and Buffalo amphibians helped the assaults across the various water obstacles and by 20 April Allied armies were racing across the Po Valley. Kesselring had always counted on the River Po as another defensive line, but Allied aircraft had destroyed all the bridges and now the river became a death trap, although some German troops managed to escape after abandoning their heavy equipment. Allied forces quickly followed up, crossing the Po and advancing towards the Alps.

Surrender
By 28 April, the Axis position was clearly hopeless and, on the 29th, von Vietinghoff, who had replaced

SICILY AND ITALY: 1943–45

AFV Units, Italy, December 1944	Commander
US Fifth Army	Lt-Gen M.L.K. Truscott
US IV Corps	Maj-Gen W.D. Crittenberger
US 1st Armoured Div	Maj-Gen V.E. Prichard
6th S African Armoured Div	Maj-Gen W. Poole
British 8th Army	Lt-Gen Sir R. McCreery
British V Corps	Lt-Gen C. Keightley
British 1st Armoured Div	Maj-Gen R. Hull
British XIII Corps	Lt-Gen S. Kirkman
British 6th Armoured Div	Maj-Gen H. Murray
Canadian I Corps	Lt-Gen C. Foulkes
Canadian 5th Armoured Div	Maj-Gen B. Hoffmeister
Polish II Corps	Lt-Gen W. Anders
Polish 2nd Armoured Bde	Brig-Gen B. Rakowski

Kesselring, signed the surrender of all German forces in Italy, which came into effect on 2 May.

Postscript

Whilst war is a grim business, it does have its lighter moments. In April 1945, Shermans of the 10th Hussars were supporting 167th Brigade's attack on Bastia, in Umbria, central Italy. One of the objectives was a thoroughly bombed treacle factory. Its damaged storage tanks had leaked thousands of litres of treacle across the surrounding area.

Fortunately, there was no opposition, as the attack quite literally bogged down in the sticky mess and the crew of one Sherman had to be rescued after their tank fell into a treacle-filled bomb crater!

▶ Wasp Mark IIC Flamethrower

Eighth Army / I Canadian Corps / 5th Canadian Armoured Division

The Wasp Mark IIC was a highly effective weapon that was widely used from August 1944 until the end of the war in Europe. On occasions Wasps were used in considerable numbers, notably in the Senio River offensive in northern Italy in April 1945, in which 127 took part.

Specifications

Crew: 3
Weight: 4.06 tonnes (4 tons)
Length: 3.76m (12ft 4in)
Width: 2.11m (6ft 11in)
Height: 1.63m (5ft 4in)
Engine: 63.4kW (85hp) Ford V8 8-cylinder petrol
Speed: 52km/h (32mph)
Range: 258km (160 miles)
Armament: 1 x flame gun

▲ Landing Vehicle Tracked (LVT-4) Buffalo IV

Eighth Army / 27th Lancers

The LVT-4 was a highly capable amphibious assault vehicle, capable of carrying troops, supplies or artillery as large as a 105mm (4.1in) howitzer. The type was extensively used in river crossings in the final stages of the Italian campaign.

Specifications

Crew: 2 plus up 35 infantry
Weight: 12.42 tonnes (12.2 tons)
Length: 7.95m (26ft 1in)
Width: 3.25m (10ft 8in)
Height: 2.46m (8ft 1in)
Engine: 186kW (250hp) Continental W670-9A 7-cylinder diesel
Speed: 32km/h (20mph)
Range: 240km (150 miles)
Armament: 1 x 12.7mm (0.5in) HMG, plus 2 x 7.62mm (0.3in) MGs

SICILY AND ITALY: 1943–45

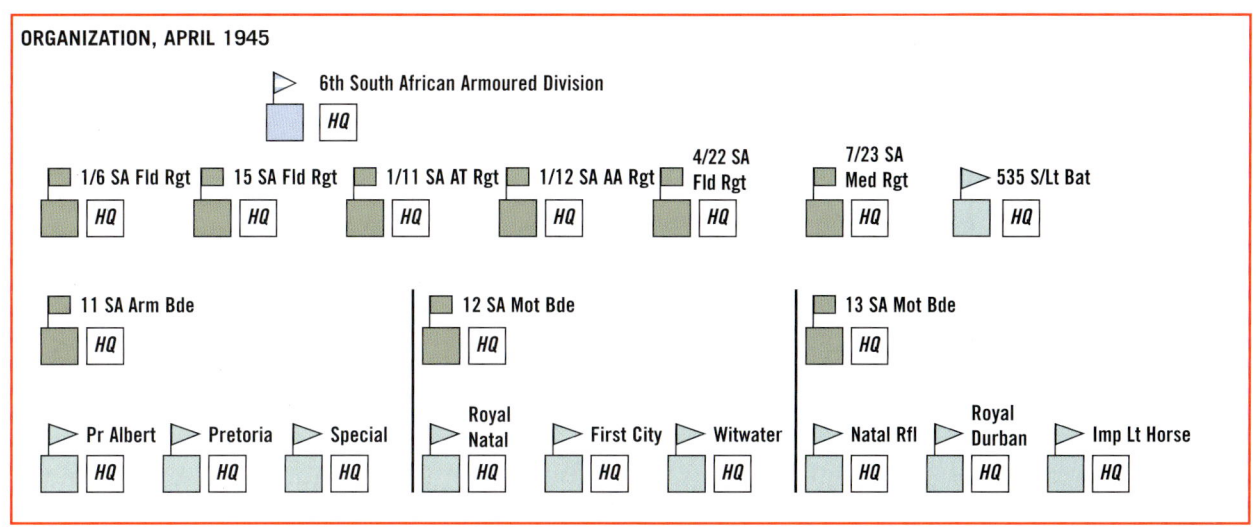

ORGANIZATION, APRIL 1945

- 6th South African Armoured Division HQ
  - 1/6 SA Fld Rgt HQ
  - 15 SA Fld Rgt HQ
  - 1/11 SA AT Rgt HQ
  - 1/12 SA AA Rgt HQ
  - 4/22 SA Fld Rgt HQ
  - 7/23 SA Med Rgt HQ
  - 535 S/Lt Bat HQ
- 11 SA Arm Bde HQ
  - Pr Albert HQ
  - Pretoria HQ
  - Special HQ
- 12 SA Mot Bde HQ
  - Royal Natal HQ
  - First City HQ
  - Witwater HQ
- 13 SA Mot Bde HQ
  - Natal Rfl HQ
  - Royal Durban HQ
  - Imp Lt Horse HQ

▲ **Taking a break**

A unit of Achilles tank destroyers and infantry in northern Italy, 1945. The powerful 17 pounder British gun was used to equip the US-supplied M10 Wolverine to produce the Achilles.

SICILY AND ITALY: 1943–45

INSIGNIA  US tank crews occasionally applied striking artwork to their vehicles, although the need for camouflage meant that it was almost always less flamboyant than that used to decorate contemporary US aircraft.

▲ M5A1 Light Tank
US Fifth Army / IV Corps / 1st Armored Division / 13th Tank Battalion

The M5A1 was the final development of the basic 'M3 design'. Although it was fast and mechanically reliable, by the end of the war it was highly vulnerable to all Axis anti-tank weapons.

Specifications

Crew: 4
Weight: 14.93 tonnes (14.7 tons)
Length: 4.34m (14ft 3in)
Width: 2.26m (7ft 5in)
Height: 2.31m (7ft 7in)
Engine: 2 x 82kW (110hp) Cadillac Series 42 V8 8-cylinder petrol
Speed: 60km/h (37mph)
Range: 161km (100 miles)
Armament: 1 x 37mm (1.5in) M6 gun, plus 3 x 7.62mm (0.3in) MGs (1 AA, 1 coaxial, 1 ball-mounted in hull front)
Radio: SCR 508

▲ M24 Chaffee Light Tank
US Fifth Army / IV Corps / 1st Armored Division / 1st Tank Battalion

The Chaffee was the most advanced light tank of the late-war period, with firepower equivalent to many of the earlier Sherman medium tanks that were still in service.

Specifications

Crew: 5
Weight: 18.28 tonnes (18 tons)
Length: 5.49m (18ft)
Width: 2.95m (9ft 8in)
Height: 2.46m (8ft 1in)
Engine: 2 x 82kW (110hp) Cadillac 44T24 V8 8-cylinder petrol
Speed: 55km/h (34mph)
Range: 282km (175 miles)
Armament: 1 x 75mm (2.9in) M6 gun, plus 1 x 12.7mm (0.5in) HMG on AA mount and 2 x 7.62mm (0.3in) MGs (1 x coaxial, 1 ball-mounted in hull front)
Radio: SCR 508

SICILY AND ITALY: 1943–45

▲ **M24 Chaffee Light Tank**
US Fifth Army / IV Corps / 1st Armored Division / 4th Tank Battalion
The M24's torsion bar suspension allowed it to maintain high cross-country speeds and its reliability equalled that of the M3/M5 series.

Specifications
Crew: 5
Weight: 18.28 tonnes (18 tons)
Length: 5.49m (18ft)
Width: 2.95m (9ft 8in)
Height: 2.46m (8ft 1in)
Engine: 2 x 82kW (110hp) Cadillac 44T24 V8 8-cylinder petrol
Speed: 55km/h (34mph)
Range: 282km (175 miles)
Armament: 1 x 75mm (2.9in) M6 gun, plus 1 x 12.7mm (0.5in) HMG on AA mount and 2 x 7.62mm (0.3in) MGs (1 x coaxial, 1 ball-mounted in hull front)
Radio: SCR 508

▲ **Up hill and down dale**
Churchills and Universal Carriers in the hills of Italy, 1945. The Churchill's superb hill-climbing ability proved to be invaluable in this sort of terrain.

Chapter 5

D-Day to Arnhem: 1944

The Allied planners responsible for the Normandy landings faced a daunting task – the disastrous Dieppe Raid had been a dire warning of what might happen if things went wrong. The sheer scale of the amphibious and airborne operations, subject to the vagaries of tide and weather conditions, posed enormous challenges. Even if these could be overcome, there remained the threat of Rommel's ingenious Atlantic Wall defences and his potentially devastating Panzer counterattacks. The 'funnies' of 79th Armoured Division provided a partial solution to the problems, but, in the final analysis, it was Allied numerical superiority and air power that proved to be the decisive factors.

◀ **AVRE**
An AVRE (Armoured Vehicle, Royal Engineers) moves up to the front-line, its turret festooned with kit.

D-DAY TO ARNHEM: 1944

79th Armoured Division

Despite their bizarre appearance, the 'funnies' of 79th Armoured Division proved to be invaluable during the long advance from Normandy to the Baltic, whether in cracking open the grandiose fortifications of the Atlantic Wall or in breaching improvised roadblocks.

IN MANY RESPECTS the Dieppe Raid of August 1942 proved to be a bloody fiasco, not least for the tanks that were landed – most of the 29 Churchills involved were unable to cross the sea wall and were knocked out on the beach. This experience convinced the British of the need for specialized AFVs, and an intensive development programme was launched that produced a vast range of such vehicles – the 'funnies' of 79th Armoured Division.

The first major step was taken in April 1943, with the appointment of Major-General Percy Hobart to command the 79th Armoured Division. His remit was to convert the division into a formation to develop and administer all specialized AFVs. Hobart was reputedly suspicious at first and would not accept command until formally assured that it would be an operational formation with a proper combat role. Under his leadership, the division rapidly developed units of extensively modified vehicles that were to become known as 'Hobart's Funnies'.

INSIGNIA

The badge of 79th Armoured Division was derived from the black bull of the family crest of Major-General Percy Hobart, the divisional commander.

Specialist troops

From the outset, it was recognized that the division would never go into action as a single formation – its

104

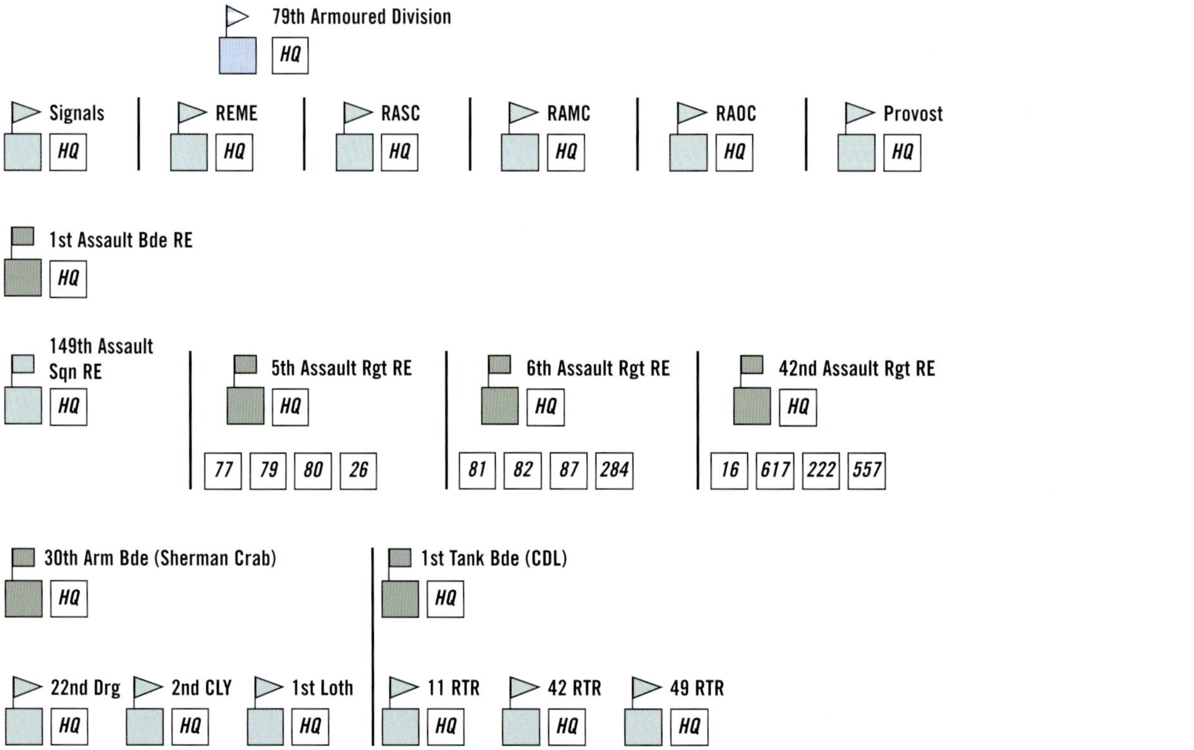

D-DAY TO ARNHEM: 1944

◀ **On target!**
A Churchill Crocodile and its blazing target, Holland, 1944.

function was to organize detachments to carry out whatever special tasks were required. Training went ahead at a furious rate, with the first major exercise at Linney Head followed by extensive 'breaching trials' against replica Atlantic Wall fortifications built at Orford training area in Suffolk. Despite spectacular demonstrations laid on for General Eisenhower and other senior American officers, US forces were

▲ **Churchill Crocodile Flamethrower**

79th Armoured Division / 31st Armoured Brigade / 1st Fife and Forfar Yeomanry

The Crocodile proved to be highly effective against all types of bunkers and pillboxes. The tank's psychological effect was such that many fortifications surrendered after the first ranging shots from a Crocodile's flame gun.

Specifications
Crew: 5
Weight: 46.5 tonnes (45.7 tons)
Length: 12.3m (40ft 6.5in)
Width: 3.2m (10ft 8in)
Height: 2.4m (8ft 2in)
Engine: 261.1kW (350hp) Bedford 12-cylinder petrol
Speed: 21km/h (13.5mph)
Range: 144km (90 miles)
Armament: 1 x 75mm (2.9in) OQF gun plus 1 x coaxial 7.92mm (0.31in) Besa MG and 1 x flame gun
Radio: Wireless Sets Nos. 19 and 38

Specifications
Crew: 5
Weight: 46.5 tonnes (45.7 tons)
Length: 12.3m (40ft 6.5in)
Width: 3.2m (10ft 8in)
Height: 2.4m (8ft 2in)
Engine: 261.1kW (350hp) Bedford 12-cylinder petrol
Speed: 21km/h (13.5mph)
Range: 144km (90 miles)
Armament: 1 x 75mm (2.9in) OQF gun, plus 1 x coaxial 7.92mm (0.31in) Besa MG and 1 x flame gun
Radio: Wireless Sets Nos. 19 and 38

▲ **Churchill Crocodile Flamethrower**

79th Armoured Division / 31st Armoured Brigade / 7 RTR

The Crocodile's armoured trailer carried a total of 1820 litres (400 gallons) of flame fuel, sufficient for up to 80 one-second shots out to a maximum range of roughly 91m (100 yards).

D-DAY TO ARNHEM: 1944

▲ **Take cover!**
A Sherman Crab, a mine-clearing variant equipped with a flail, comes under fire as the Allies push through Normandy, July 1944.

▲ **Sherman Crab Flail Tank**
79th Armoured Division / 30th Armoured Brigade / 22nd Dragoons

This Crab, with guns and mantlet covered for transit, is unusual in retaining the 12.7mm (0.5in) Browning heavy machine gun on the turret roof. The odd-looking 'stalks' on the hull and turret held station-keeping lights for use when several Crabs were flailing together, throwing up clouds of dust and debris. In such poor visibility careful control was essential to ensure that the tanks didn't drift apart, leaving an uncleared strip of ground between them.

Specifications

Crew: 5
Weight: 31.8 tonnes (31.3 tonnes)
Length: 8.23m (27ft)
Width: 3.5m (11ft 6in)
Height: 2.7m (9ft)
Engine: 373kW (500hp) Ford GAA V8 petrol
Speed: 46km/h (29mph)
Range: 100km (62 miles)
Armament: 1 x 75mm (2.9in) M3 gun, plus 1 x 12.7mm (0.5in) AA HMG and 1 x 7.62mm (0.3in) MG
Radio: Wireless Set No. 19

D-DAY TO ARNHEM: 1944

sceptical of the ability of the strange-looking 'funnies' to operate effectively on the battlefield and refused to adopt any of them except for the amphibious Sherman DD.

Battlefield experience was to show that this was a dire mistake and after the Normandy landings detachments of 'funnies' were in great demand to support US assaults on heavily fortified positions such as the Channel ports. By the end of the war the division had 21,430 men and 1566 AFVs – a conventional armoured division's strength was 14,400 men and 350 AFVs.

The vast majority of the division's AFVs were modified versions of the Churchill or the Sherman,

Specifications
Crew: 5
Weight: 31.8 tonnes (31.3 tonnes)
Length: 8.23m (27ft)
Width: 3.5m (11ft 6in)
Height: 2.7m (9ft)
Engine: 373kW (500hp) Ford GAA V8 petrol
Speed: 46km/h (29mph)
Range: 100km (62 miles)
Armament: 1 x 75mm (2.9in) M3 gun, plus 1 x 7.62mm (0.3in) MG
Radio: n/k

▲ **Sherman Crab Flail Tank**
79th Armoured Division / 30th Armoured Brigade / 2nd County of London Yeomanry (Westminster Dragoons)
The serrated edges of the flail drum acted as wire cutters, preventing the flail chains becoming entangled in barbed-wire obstacles.

▲ **Sherman Crab Flail Tank**
79th Armoured Division / 30th Armoured Brigade / 1st Lothian and Border Horse Yeomanry
A variety of special equipment was added to Crabs as a result of operational experience, including side-mounted angled boxes containing powdered chalk that trickled out to mark the swept lane.

Specifications
Crew: 5
Weight: 31.8 tonnes (31.3 tonnes)
Length: 8.23m (27ft)
Width: 3.5m (11ft 6in)
Height: 2.7m (9ft)
Engine: 373kW (500hp) Ford GAA V8 petrol
Speed: 46km/h (29mph)
Range: 100km (62 miles)
Armament: 1 x 75mm (2.9in) M3 gun, plus 1 x 7.62mm (0.3in) MG
Radio: n/k

D-DAY TO ARNHEM: 1944

both of which were readily available in large numbers. Although relatively slow, the Churchill had a roomy interior, thick armour and a good cross-country performance. Whilst the Sherman was less well armoured, its greater speed and exceptional mechanical reliability made it particularly suitable for other roles undertaken by the formation.

Types of 'funnies'

Churchill and Sherman tanks were adapted to perform a multitude of battlefield tasks beyond the capabilities of conventional AFVs. Among the many 'funnies' developed by Hobart's division were:

• Churchill Crocodile – The first Crocodiles were converted from Churchill IVs, but production models were all based on the more heavily armoured Churchill VII. This conversion involved fitting the flame gun of the Wasp Mark II Flamethrower in the Churchill's hull machine-gun mount and installing piping for the flame fuel which was run back to the 'Link', a three-way coupling on the hull rear plate. This was the connection to the armoured trailer that contained 1820 litres (400 gallons) of flame fuel,

▲ **Up and over!**
An AVRE practising obstacle-crossing using fascines.

▲ **Churchill AVRE**
79th Armoured Division / 1st Assault Brigade RE / 6th Assault Regiment RE
AVREs landed in the first assault waves were carefully waterproofed and fitted with deep-wading equipment, including trunking over the air intakes and extended exhausts. This example mounts a 'Bobbin Carpet' – a reinforced mat that was laid over soft sand to provide a temporary trackway for following vehicles.

Specifications (without attachments)
Crew: 6
Weight: 38.6 tonnes (38 tons)
Length: 7.67m (25ft 2in)
Width: 3.25m (10ft 8in)
Height: 2.79m (9ft 2in)
Engine: 261.1kW (350hp) Bedford 12-cylinder petrol
Speed: 25km/h (15.5mph)
Range: 193km (120 miles)
Armament: 1 x 290mm (11.4in) Petard demolition mortar, plus 2 x 7.92mm (0.31in) Besa MGs (1 coaxial and 1 ball-mounted in hull front)
Radio: Wireless Set No. 19

D-DAY TO ARNHEM: 1944

▼ 79th Armoured Division – Assault Detachment

The 'funnies' were constantly on the move, forming and re-forming detachments tailored to meet the varying needs of specific operations. This force is representative of a typical grouping for an assault on a fortified sector of the German front-line.

Troop 1 (3 x Sherman Crabs) **Troop 2 (3 x Sherman Crabs)**

Troop 1 (3 x Churchill Crocodiles)

Troop 2 (3 x Churchill Crocodiles)

Troop 3 (3 x Churchill Crocodiles)

Troop 1 (3 x AVRE) **Troop 2 (3 x AVRE)**

Troop 3 (3 x AVRE)

Troop (3 x ARK)

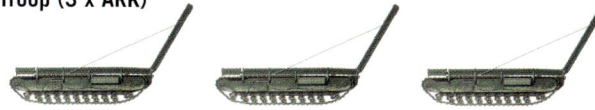

Troop (12 x Ram Kangaroos)

sufficient for 80 one-second bursts. The trailer, which proved unexpectedly resistant to battle damage, also carried five nitrogen cylinders that pressurized the fuel. A total of 800 Crocodiles were eventually produced.

- Churchill Armoured Vehicle, Royal Engineers (AVRE) – Several hundred Churchill IIIs and IVs were converted to AVREs by having their turret armament replaced with the Petard, a 290mm (11.4in) spigot mortar. This had a practical range of 73m (80 yards) with its standard ammunition, a fin-stabilized 18kg (40lb) hollow-charge bomb that proved highly effective against a wide range of obstacles and fortifications. (An attempt to improve

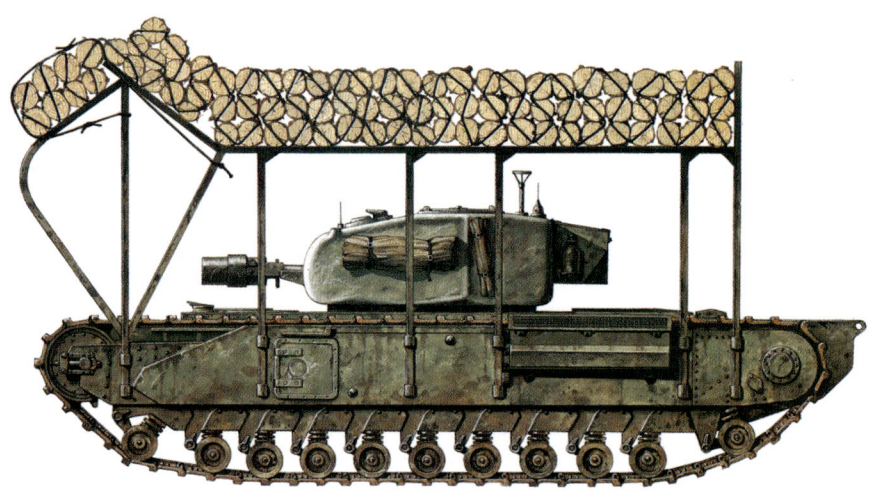

▲ **Churchill AVRE**
79th Armoured Division / 1st Assault Brigade RE / 5th Assault Regiment RE
This AVRE carries a 'Log Carpet', a more durable version of the 'Bobbin Carpet'.

Specifications (without attachments)
Crew: 6
Weight: 38.6 tonnes (38 tons)
Length: 7.67m (25ft 2in)
Width: 3.25m (10ft 8in)
Height: 2.79m (9ft 2in)
Engine: 261.1kW (350hp) Bedford 12-cylinder petrol
Speed: 25km/h (15.5mph)
Range: 193km (120 miles)
Armament: 1 x 290mm (11.4in) Petard demolition mortar, plus 2 x 7.92mm (0.31in) Besa MGs (1 coaxial and 1 ball-mounted in hull front)
Radio: Wireless Set No. 19

Specifications (without attachments)
Crew: 6
Weight: 38.6 tonnes (38 tons)
Length: 7.67m (25ft 2in)
Width: 3.25m (10ft 8in)
Height: 2.79m (9ft 2in)
Engine: 261.1kW (350hp) Bedford 12-cylinder petrol
Speed: 25km/h (15.5mph)
Range: 193km (120 miles)
Armament: 1 x 290mm (11.4in) Petard demolition mortar, plus 2 x 7.92mm (0.31in) Besa MGs (1 coaxial and 1 ball-mounted in hull front)
Radio: Wireless Set No. 19

▲ **Churchill AVRE**
79th Armoured Division / 1st Assault Brigade RE / 42nd Assault Regiment RE
Another AVRE in full deep-wading kit, fitted with a Bullshorn mine plough and towing a Porpoise skid trailer. These skid trailers were designed to carry a wide variety of supplies.

D-DAY TO ARNHEM: 1944

on the Petard resulted in the conversion of a single AVRE to mount 'Ardeer Aggie', a large-calibre recoilless demolition gun. This weapon operated on the principle of the World War I Davis Gun, which was developed for use on aircraft and in which recoil was eliminated by firing a counterweight to the rear. Ardeer Aggie used sand as a counterweight, but trials showed that the idea was impractical – the risk of sand-blasting the AVRE's engine decks or nearby infantry with every shot was just too great.

On a more practical level, AVREs were adapted to carry and operate a wide range of special equipment such as:

– The Bobbin: A reel of 3m (10ft) wide canvas cloth reinforced with steel poles carried above the tank and unrolled onto patches of soft ground to form a 'carpet' to prevent vehicles bogging down.

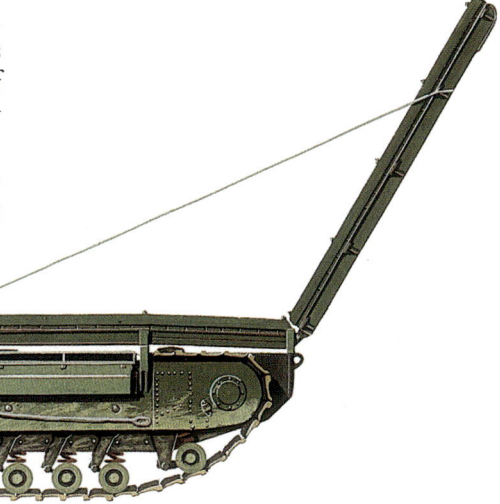

▲ **Churchill ARK Mark I**
79th Armoured Division / 1st Assault Brigade RE / 5th Assault Regiment RE
ARKs were highly effective – in one case, two were used, one on top of the other, to bridge a deep ravine in Italy.

Specifications
Crew: 2/3
Weight: 35 tonnes (34.44 tons) (estimated)
Length: 7.67m (25ft 2in)
Width: 3.25m (10ft 8in)
Height: 1.83m (6ft) (estimated)
Engine: 261.1kW (350hp) Bedford 12-cylinder petrol
Speed: 25km/h (15.5mph)
Range: 193km (120 miles)
Armament: 1 or 2 x 7.7mm (0.303in) Bren MGs
Radio: Wireless Set No. 19

▲ **Churchill AVRE**
79th Armoured Division / 1st Assault Brigade RE / 6th Assault Regiment RE
This AVRE carries one fascine and tows another on an AVRE skid trailer. Fascines were simple and highly effective obstacle-crossing devices, but badly obscured the driver's vision. In most cases another crew member had to perch on top of the fascine to guide the driver.

Specifications (without attachments)
Crew: 6
Weight: 38.6 tonnes (38 tons)
Length: 7.67m (25ft 2in)
Width: 3.25m (10ft 8in)
Height: 2.79m (9ft 2in)
Engine: 261.1kW (350hp) Bedford 12-cylinder petrol
Speed: 25km/h (15.5mph)
Range: 193km (120 miles)
Armament: 1 x 290mm (11.4in) Petard demolition mortar, plus 2 x 7.92mm (0.31in) Besa MGs (1 coaxial and 1 ball-mounted in hull front)
Radio: Wireless Set No. 19

D-DAY TO ARNHEM: 1944

– Fascines: Bundles of wooden poles or brushwood lashed together with wire. These were carried on the front of the tank and could be released to fill a ditch or form a step. Metal pipes in the centre of the fascine allowed water to flow through.

– Small Box Girder (SBG) Assault Bridge: A bridge carried on the front of the tank, which could be dropped to span a 9.1m (30ft) gap in 30 seconds.

– Bullshorn Plough: This mine plough proved to be effective on soft ground, lifting and turning aside mines for subsequent defusing.

Specifications
Crew: 5
Weight: 32.3 tonnes (31.8 tons)
Length: 6.35m (20ft 10in)
Width: 2.81m (9ft 3in)
Height: 3.96m (13ft)
Engine: 373kW (500hp) Ford GAA V8 petrol
Speed (water): 7.4km/h (4 knots)
Range: 240km (149 miles)
Armament: 1 x 75mm (2.9in) M3 gun, 1 x coaxial 7.62mm (0.3in) MG
Radio: Wireless Set No. 19

▶ **Sherman DD**
3rd Infantry Division / 27th Armoured Brigade / 4th/7th Royal Dragoon Guards
Although the waterproofed canvas flotation screen of the Sherman DD looked fragile, operational experience showed that it could withstand surprisingly rough sea conditions.

Specifications
Crew: 5
Weight: 32.3 tonnes (31.8 tons)
Length: 6.35m (20ft 10in)
Width: 2.81m (9ft 3in)
Height: 3.96m (13ft)
Engine: 373kW (500hp) Ford GAA V8 petrol
Speed (water): 7.4km/h (4 knots)
Range: 240km (149 miles)
Armament: 1 x 75mm (2.9in) M3 gun, 1 x coaxial 7.62mm (0.3in) MG
Radio: Wireless Set No. 19

▶ **Sherman DD**
3rd Infantry Division / 27th Armoured Brigade / 13th/18th Royal Hussars
Although best known for their role in the Normandy landings, Sherman DD tanks were also used in the assault crossings of the Rhine and Elbe.

D-DAY TO ARNHEM: 1944

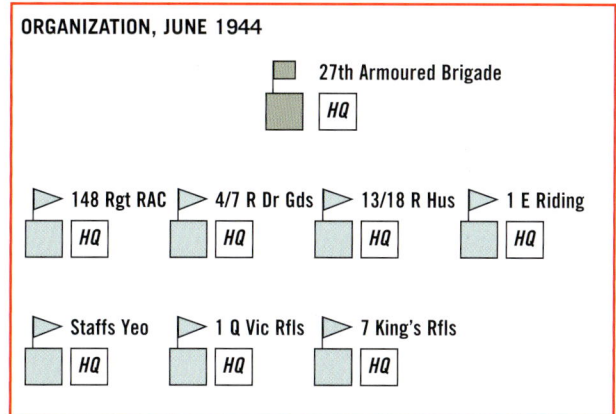

ORGANIZATION, JUNE 1944

27th Armoured Brigade HQ

148 Rgt RAC | 4/7 R Dr Gds | 13/18 R Hus | 1 E Riding

Staffs Yeo | 1 Q Vic Rfls | 7 King's Rfls

– Double Onion: Two large demolition charges on a metal frame that could be placed against a concrete wall and then detonated from a safe distance. It was the successor to the single-charge devices Carrot and Light Carrot, which were trialled but never used in combat.

• Churchill Armoured Ramp Carrier (ARK) – A number of Churchills were converted to ARKs by the removal of their turrets and the fitting of moveable ramps at each end. When the tank was driven up to a sea wall (or into a crater or gulley), the ramps were deployed, allowing other vehicles to drive up the first ramp, over the ARK, then along the second ramp, enabling the obstacle to be scaled.

▲ **DD column**
Sherman DD tanks move along a road in Normandy shortly after the Allied invasion, June 1944. The DD tanks would again prove their usefulness in the Rhine and Elbe crossings of 1945.

Specifications
Crew: 4
Weight: 28.45 tonnes (28 tons)
Length: 5.63m (18ft 5in)
Width: 2.58m (8ft 6in)
Height: 2.83m (9ft 2in)
Engine: 305.45kw (410hp) General Motors 6046 12-cylinder twin in-line diesel
Speed: 48km/h (30mph)
Range: 240km (150 miles)
Radio: Wireless Set No. 19

▲ **Sherman BARV**
79th Armoured Division / REME
Roughly 60 Sherman BARVs were used on the invasion beaches of Normandy. Unusually for a tank, the crew included a diver, whose job was to attach towing gear to 'drowned' vehicles.

113

- Sherman Crab – A modified Sherman tank which was equipped with a mine-clearing flail: a rotating drum fitted with weighted chains that beat the ground to detonate mines in the path of the tank.
- Sherman DD – From 'Duplex Drive', this was an amphibious Sherman able to swim ashore after launching from a landing craft well offshore to give support to the first waves of infantry immediately on landing. (Similarly adapted Valentines were used only for trials and training.)
- Sherman Beach Armoured Recovery Vehicle (BARV) – These were M4A2 Shermans that were waterproofed and had their turrets replaced with tall armoured superstructures. These tanks were capable of operating in up to 2.7m (9ft) of water to remove broken-down or swamped vehicles that were blocking access to the beaches. BARVs were also used to refloat small landing craft that had become stuck on the beach.
- Buffalo – The British version of the American LVT-4 amphibious landing vehicle, the Buffalo was frequently fitted with appliqué armour and up-gunned, with some examples carrying a forward-

▶ **'Hobo'**
Major-General Percy Hobart, commander of the 79th Armoured Division and the driving force behind the creation of the 'funnies'.

▲ **Tank recovery**
A Cromwell ARV (Armoured Recovery Vehicle) tows a captured Panzer Mk IV. The Cromwell ARV was a turretless conversion of a Cromwell tank.

D-DAY TO ARNHEM: 1944

firing Browning machine gun in a ball mount operated by the co-driver, a 20mm (0.79in) Polsten cannon on the cab roof and a further Browning machine gun on each side of the cargo bay.

- Armoured Bulldozer – This 'funny' was not a converted tank but a standard Caterpillar D8 Bulldozer equipped with armour protection for the driver and the engine. The vehicle's primary role was to clear the invasion beaches of obstacles and to make roads accessible by clearing rubble and filling in bomb craters. Conversions were carried out by a Caterpillar importer: Jack Olding & Co Ltd of Hatfield, Hertfordshire.

- Centaur Bulldozer – An obsolescent Centaur tank with the turret removed and fitted with a simple, winch-operated, bulldozer blade, the Centaur

Specifications
Crew: 5
Weight: 27.22 tonnes (26.7 tons) (estimated)
Length: 5.64m (18ft 6in)
Width: 2.72m (8ft 11in)
Height: 3.12m (10ft 3in) (esimated)
Engine: 253.5kW (340hp) Continental R-975-EC2 radial petrol
Speed: 42km/h (26mph)
Range: 193km (120 miles)
Armament: 1 x turret-mounted searchlight, 1 x 75mm (2.9in) M2 or M3 gun, 1 x 7.92mm (0.31in) Besa MG
Radio: Wireless Set No. 19

▲ **Grant Canal Defence Light (CDL)**
79th Armoured Division / 1st Tank Brigade / 11 RTR
The CDL was an ingenious device that suffered from excessive security – in effect, it was so secret that almost no one knew of it and so it was scarcely ever used. However, a detachment was used during the Rhine crossings, illuminating and destroying a number of floating mines launched against bridges and ferries.

Specifications
Crew: 2 plus up 35 infantry
Weight: 12.42 tonnes (12.2 tons)
Length: 7.95m (26ft 1in)
Width: 3.25m (10ft 8in)
Height: 2.46m (8ft 1in)
Engine: 186kW (250hp) Continental W670-9A 7-cylinder
Speed: 32km/h (20mph)
Range: 240km (150 miles)
Armament: 1 x 12.7mm (0.5in) HMG, plus 2 x 7.62mm (0.3in) MGs

▲ **Landing Vehicle Tracked (LVT) 2 'Water Buffalo'**
79th Armoured Division / 33rd Armoured Brigade / 1st Northamptonshire Yeomanry
The Buffalo's 'main armament' of a 12.7mm (0.5in) heavy machine gun was frequently replaced with a 20mm (0.79in) cannon.

D-DAY TO ARNHEM: 1944

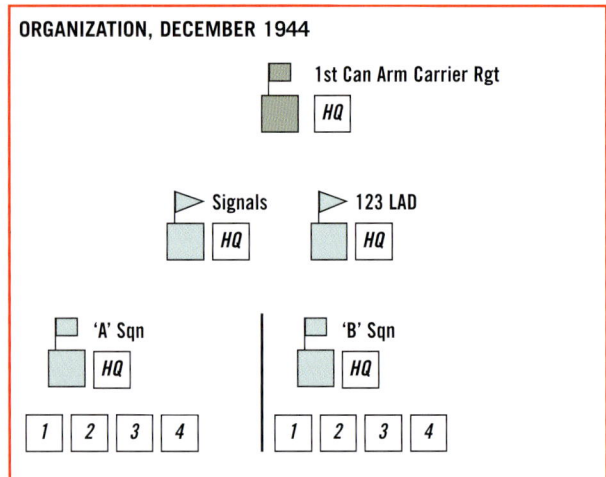

ORGANIZATION, DECEMBER 1944

Bulldozer was produced because of a need for a well-armoured, obstacle-clearing vehicle that, unlike a conventional bulldozer, was fast enough to keep up with tank formations. These vehicles were not used on D-Day but were issued to the 79th Armoured Division in Belgium during the latter part of 1944.

• Grant Canal Defence Light (CDL) – This was a powerful carbon-arc searchlight in a special turret fitted to a modified Grant. The device was given a deliberately misleading name for security reasons – its true purpose was to blind the defenders during night attacks. An ingenious optical design allowed the light to flood through a small slit in the armour, minimizing the risk of damage by enemy fire. The type was not used on D-Day, but was employed to create 'artificial moonlight' during the later attack on the Geilenkirchen Salient.

• Kangaroos – These armoured personnel carriers were devised by II Canadian Corps to minimize future infantry casualties after the heavy losses incurred in the early fighting to expand the Normandy beachheads. The first Kangaroos were converted from a batch of 102 surplus Priest self-propelled howitzers taken from artillery units that had re-equipped with 25pdr guns. At a field workshop (codenamed Kangaroo, hence the name), they were stripped of their howitzers, the gun mounting was plated over and bench seats for 12 infantrymen were fitted. They were first used in Operation *Totalize*, before being supplemented by further variants converted from Rams and Shermans which were widely used for the remainder of the war.

▲ Better than marching
A Ram Kangaroo carrying its infantry squad, autumn 1944.

D-DAY TO ARNHEM: 1944

▼ 1st Canadian Armoured Carrier Regiment, 'A' Squadron

Whilst several kinds of AFV were converted to Kangaroo APCs, the obsolescent Canadian Ram cruiser tank was the commonest base type. Ram Kangaroos had the advantage of similar levels of protection and mobility to the Shermans with which they most frequently operated, although their open-topped troop compartments were vulnerable to air-burst artillery fire. Initially they were operated by 1st Canadian Armoured Personnel Carrier Squadron, which was soon expanded to become 1st Canadian Armoured Carrier Regiment within 79th Armoured Division. By late 1944, it had been joined by the British 49th Armoured Personnel Carrier Regiment. Each unit fielded a total of 106 Ram Kangaroos and both were heavily committed to operations in support of Twenty-first Army Group until VE Day.

1 Troop

2 Troop

3 Troop

4 Troop

▲ Ram Kangaroo
79th Armoured Division / 31st Armoured Brigade / 1st Canadian Armoured Carrier Regiment

Several types of tanks and self-propelled guns, including Rams, Priests and Shermans, were converted to APCs (all of which were known as Kangaroos) by removing their main armament and fitting bench seats in the fighting compartment. The maximum number of passengers varied depending on the vehicle type, and armament varied wildly – all types mounted at least one machine gun, but extra pintle-mounted machine guns, heavy machine guns and 20mm (0.79in) cannon were 'acquired' whenever possible.

Specifications
Crew: 2 plus 10 infantry
Weight: 25.4 tonnes (25 tons)
Length: 5.79m (19ft)
Width: 2.77m (9ft 1in)
Height: 1.91m (6ft 3in)
Engine: 298kW (400hp) Continental R-975 9-cylinder radial petrol
Speed: 40km/h (25mph)
Range: 232km (144 miles)
Armament: 1 x 7.62mm (0.3in) MG in sub-turret
Radio: Wireless Set No. 19

D-DAY TO ARNHEM: 1944

US armoured engineer AFVs 1944–45

US armoured engineering suffered from the lack of an equivalent of the 79th Armoured Division. Some work was carried out, particularly on mine-clearing devices and AFV-mounted flamethrowers, but most US formations came to rely on detachments of 'funnies' for armoured engineering tasks.

THE MAIN US MINE-EXPLODER device to see service, the T1E3, was named 'Aunt Jemima' after a popular pancake mix logo, as the big exploder wheels resembled gigantic pancakes. 'Jemima' comprised two massive steel rollers pushed in front of a Sherman, with one roller ahead of each tank track. Each roller was divided into five discs, each of which was about 10cm (3.9in) thick and 3m (10ft) in diameter. The whole device weighed in at almost 27,000kg (60,000lb), theoretically more than enough to detonate any mine.

Roller chains from the Sherman's sprockets drove the loosely mounted discs, and the T1E3 was found to work well in tests. In service, however, it was a different story – the device proved to be difficult to manoeuvre and its immense weight caused it to bog down repeatedly. As many as 70 'Jemimas' may have been issued, but US units heartily loathed them, much preferring to call on 79th Armoured Division's Sherman Crabs.

US Armoured Engineer Battalion	Strength
HQ Company:	
¼ ton truck	2
¾ ton truck	2
M3 halftracks	3
2½ ton truck	1
1 ton trailer	1
Company HQ:	
¼ ton truck	8
¾ ton truck	2
2½ ton truck	14
1 ton trailer	11
6 ton heavy wrecker	1
Motorised shop	1
Welding equipment trailer	1
Truck mounted compressor	1
3 ton bridge truck	3
water equipment trailer	4
M10 ammo trailer	2
Armoured Engineer Company:	
¼ ton truck	4
¾ ton truck	1
M3 halftrack	4
2½ ton truck	3
1 ton trailer	3
Truck mounted compressor	1
Tractor (bulldozer)	1
20 ton semi-trailer and tractor	1
6 ton bridge truck	1
2½ ton utility trailer	2
2½ ton dump truck	3

▲ **T15E1 Mine Resistant Vehicle**
Unidentified trials unit, continental USA
Whilst the idea of a mine resistant tank seemed attractive, it proved to be hopelessly impractical. Besides the difficulty of building a vehicle strong enough to cope with the repeated explosions of anti-tank mines, there was the problem of the 'unswept' strip of ground between the tracks.

Specifications
Crew: 5
Weight: 32.2 tonnes (31.6 tons)
Length: 5.9m (19ft 4in)
Width: 2.75m (9ft)
Height: 2.04m (6ft 8in)
Engine: 2 x 373kW (500hp) General Motors 6-cylinder petrol
Speed: 48km/h (30mph)
Range: 270km (170 miles)
Radio: n/k

D-DAY TO ARNHEM: 1944

Flamethrowing tanks

US forces converted significant numbers of Stuarts and Shermans as armoured flamethrowers, but these were mainly deployed in the Pacific and South-East Asia. Four Shermans were converted (probably by 79th Armoured Division's workshops) to M4 Crocodiles, using standard Churchill Crocodile armoured fuel trailers but with an armoured fuel line running over the hull to a flame gun mounted next to the hull gunner's hatch. It is uncertain if any of these vehicles ever went into action, although they did serve with the US Ninth Army – in general, detachments of Churchill Crocodiles provided flamethrower support for US units.

▲ **Mine Exploder T1E3 (M1) 'Aunt Jemima'**
US Ninth Army / 739th Tank Battalion
The 'Aunt Jemima' was even trickier to manoeuvre than the Sherman Crab and was never popular with US armoured units.

Specifications
Crew: 5
Weight: 57 tonnes (56.1 tons) (estimated)
Length: 9.9m (32ft 6in) (estimated)
Width: 2.82m (9ft 3in) (estimated)
Height: 3m (9ft 10in)
Engine: 372.5kW (500hp) Ford GAA 8-cylinder petrol
Speed: 24km/h (15mph) (estimated)
Range: 110km (68.3 miles)
Armament: 1 x 75mm (2.9in) M3 gun, plus 1 x coaxial 7.62mm (0.3in) MG
Radio: SCR508

▲ **M4 Crocodile**
US Ninth Army / 739th Tank Battalion
The Sherman's thinner armour made it less well suited to the armoured flamethrower role than the Churchill. Only a single platoon of four M4 Crocodiles became operational with the 739th Tank Battalion from November 1944.

Specifications (tank alone)
Crew: 5
Weight: 30.3 tonnes (29.82 tons)
Length: 5.90m (19ft 4in)
Width: 2.62m (8ft 7in)
Height: 2.74m (8ft 11in)
Engine: 372.5kW (500hp) Ford GAA 8-cylinder petrol
Speed: 42km/h (26mph)
Range: 210km (130 miles)
Armament: 1 x 75mm (2.9in) M3 gun, 1 x flame gun by co-driver's hatch and 2 x 7.62mm (0.3in) MGs (1 coaxial, 1 ball-mounted in hull front)
Radio: SCR 508

D-DAY TO ARNHEM: 1944

Normandy to Arnhem
JUNE–SEPTEMBER 1944

The three and a half months between the Normandy landings and the end of Operation *Market Garden* were marked by grinding attritional warfare, followed by exhilarating pursuits and finally the frustration of a 'lost victory'. Throughout it all, Allied armour played a key role in countless actions, from great offensives to minor skirmishes.

IN THE MONTHS before the Normandy landings, reports from armoured units in Italy began to raise suspicions that the Sherman was rapidly becoming outclassed by the newer German AFVs such as the Panther and Jagdpanzer IV. American experts tended to ignore such concerns – the official US Army view was that updated versions of the Sherman with the new 76mm (3in) gun would be quite capable of holding their own against any German tank. As a result, the enormous American tank production effort remained concentrated on improved versions of the Sherman until the last stages of the war in Europe, when it was belatedly accepted that new types were desperately needed to match the increasingly sophisticated German AFVs.

British designs were virtually a generation behind their German counterparts, but the first issues of tungsten-cored APDS (armour-piercing, discarding

▼ **Reinforcements**
Columns of Shermans and Churchills pass each other on a country road 'somewhere in France'.

D-DAY TO ARNHEM: 1944

sabot) ammunition for the 6pdr and 17pdr in the summer of 1944 marked a major advance in anti-tank technology. Fortunately, the new APDS round gave the 17pdr an anti-tank performance at least equal to all German AFV weapons, except for the rarely encountered 128mm (5in) of the Jagdtiger.

Producing the weapon was one thing, but finding suitable AFVs to mount it was quite another. Luckily, the 17pdr could be squeezed into a modified Sherman turret and that of the M10 tank destroyer, which were backed up by 650-plus Archers and a small number of Challengers. Unfortunately, there

▲ M4A3 Sherman Medium Tank
US First Army / V Corps / 2nd Armored Division / 66th Armored Regiment

This Sherman's exhaust and air intakes are fitted with deep-wading trunking. Many vehicles had to be landed on the original invasion beaches long after D-Day and needed waterproofing or deep-wading equipment to minimize the risk of being swamped as they came ashore.

Specifications
Crew: 5
Weight: 30.3 tonnes (29.82 tons)
Length: 5.90m (19ft 4in)
Width: 2.62m (8ft 7in)
Height: 2.74m (8ft 11in)
Engine: 372.5kW (500hp) Ford GAA 8-cylinder petrol
Speed: 42km/h (26mph)
Range: 210km (130 miles)
Armament: 1 x 75mm (2.9in) M3 gun, plus 2 x 7.62mm (0.3in) MGs (1 coaxial, 1 ball-mounted in hull front)
Radio: SCR508

▲ 75mm Howitzer Motor Carriage (HMC) M8
US First Army / V Corps / 2nd Armored Division / 24th Cavalry Reconnaissance Squadron

Based on the chassis of the M5 (Stuart), the M8 had an open-topped turret with a 75mm (2.9in) M2 or M3 howitzer. Most M8s were deployed to provide fire support to the M5s of reconnaissance squadrons. This vehicle is fitted with a 'Culin' hedgerow cutter for breaking through the thick hedges of the bocage.

Specifications
Crew: 4
Weight: 15.6 tonnes (15.45 tons)
Length: 4.41m (14ft 6in)
Width: 2.24m (7ft 4in)
Height: 2.32m (7ft 7in)
Engine: 2 x 81.95kW (110hp) Cadillac Series 42 V8 petrol
Speed: 56km/h (35mph)
Range: 210km (130 miles)
Armament: 1 x 75mm (2.9in) M2 howitzer, plus 1 x 12.7mm (0.5in) AA HMG
Radio: SCR508

D-DAY TO ARNHEM: 1944

▲ **Traffic control**
A Polish Sherman is directed through the ruins of Caen.

to the war in the East, the occupied territories of the West had come to be regarded as 'cushy' postings where German units could re-form and recuperate before returning to the Eastern Front.

Apart from training formations, very few armoured units were permanently stationed in these areas until the threat of invasion became acute in the spring of 1944. Despite the ever more urgent demands for Panzer units to stem the increasingly dangerous Soviet advances, elite formations, including 1st and 12th SS Panzer Divisions and *Panzer Lehr*, were moved to positions within striking distance of the likely invasion beaches on the Channel coast. Rommel, who had been appointed to command Army Group B covering those very beaches in December 1943, had drastically reinforced their defences.

Over two million mines were laid in the six months before D-Day, bringing the total to roughly four million, whilst 500,000 anti-tank and anti-glider obstacles blocked likely landing grounds. Rommel wanted to do more, planning to create a continuous minefield 1000m (1094 yards) deep along the whole Channel coast. This would be defended by all the available Panzer units, with the aim of destroying the invasion on the beaches. The Panzers would thus be

were never enough AFVs capable of mounting the 17pdr, and those that were available had to be doled out in small numbers.

German armour

In the aftermath of the stunning German victories of 1940, France, Belgium and Holland became something of a backwater. By the end of 1941, when it became clear that there would be no quick ending

Specifications
Crew: 3
Weight: 7.62 tonnes (7.5 tons)
Length: 4.11m (13ft 6in)
Width: 2.31m (7ft 7in)
Height: 2.12m (6ft 11.5in)
Engine: 123kW (165hp) Meadows MAT 12-cylinder petrol
Speed: 64km/h (40mph)
Range: 225km (140 miles)
Armament: 1 x 76mm (3in) CS howitzer, plus 1 x coaxial 7.92mm (0.31in) Besa MG
Radio: Wireless Set No. 19

▲ **Tetrarch ICS**

6th Airborne Division / Airborne Armoured Reconnaissance Regiment
The Light Tank Mark VII entered production in 1940 and was adopted by the airborne forces in 1941 as it was light enough to be carried by the new Hamilcar glider. A small number were flown in to 6th Airborne Division's landing zones early on D-Day. They provided close fire support for the lightly equipped airborne infantry for a few days until relieved by armoured units advancing from the beaches. Most Tetrarchs were armed with the 2pdr; this example is one of a small number of Close Support (CS) variants that carried the 76mm (3in) CS howitzer.

D-DAY TO ARNHEM: 1944

close enough to attack when the Allies were at their most vulnerable in the first few hours after landing, which would cut down the time the tanks were exposed to air attacks and naval gunfire. His ideas on Panzer deployment fell foul of his immediate superior, the Commander-in-Chief West, von Runstedt, who wanted to hold the tanks in reserve before launching them in a major counter-offensive once the main Allied attack had been slowed by the infantry divisions defending the Channel coast. This dispute was finally settled in a compromise – three of the six Panzer divisions were assigned to Rommel whilst the remainder, designated Panzer Group West, were held back under Hitler's control.

D-Day

British and American planners devised markedly different tactics for the use of armour on D-Day (6 June 1944). The US mistrust of the 'funnies' was such that they only deployed DD Shermans (many of which sank in the rough seas after being launched too far offshore). Their loss left the assault infantry and engineers to take heavy casualties from the defences, and they would have been horribly vulnerable if the Panzers had been close enough to intervene in the first crucial hours. In marked contrast, the British and Canadian landings were supported by all types of

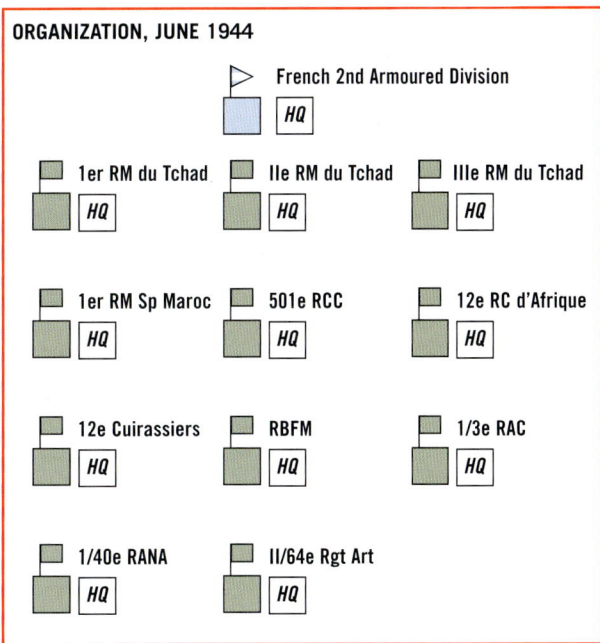

'funnies', which undoubtedly saved many lives. They also made a major contribution to the rate of advance. By nightfall on D-Day, leading British and Canadian units were almost 10km (6 miles) inland, whilst US forces were no more than 5km (3 miles) inland and had sustained a far higher casualty rate.

▲ **M4A2 Sherman Medium Tank**
US Third Army / V Corps / French 2nd Armoured Division (2e Division Blindée, 2e DB) / 12e Regiment de Chasseurs d'Afrique
2e DB took part in the Allied breakout from Normandy, in which it served as a link between US and Canadian forces. It also played a key role in the virtual destruction of 9th Panzer Division in the fighting to eliminate the Falaise Pocket.

Specifications
Crew: 5
Weight: 31.8 tonnes (31.29 tons)
Length: 5.92m (19ft 5in)
Width: 2.62m (8ft 7in)
Height: 2.74m (8ft 11in)
Engine: 305.45kW (410hp) General Motors 6046 12-cylinder twin in-line diesel
Speed: 48km/h (30mph)
Range: 240km (150 miles)
Armament: 1 x 75mm (2.9in) M3 gun, plus 2 x 7.62mm (0.3in) MGs (1 coaxial, 1 ball-mounted in hull front)
Radio: SCR508

D-DAY TO ARNHEM: 1944

▲ **SP 17pdr, Valentine Mk I, Archer**
Twenty-first Army Group / Second Army / VIII Corps / 3rd Infantry Division / 20th Anti-Tank Regiment RA

The bulk and weight of the very effective 17pdr anti-tank gun were such that the provision of self-propelled versions was a matter of priority. Rearming the US-supplied M10 Wolverine tank destroyers provided a partial solution, but there were never enough to meet demand. Fortunately, the Valentine was still in production and it was decided to use its hull as the basis for a new vehicle, designated Archer. The 17pdr was carried on a rearward-facing, limited-traverse mount in an open-topped fighting compartment and the result was surprisingly successful, having a low silhouette and good agility.

Specifications

Crew: 4
Weight: 18.79 tonnes (18.5 tons)
Length: 6.68m (21ft 11in)
Width: 2.64m (8ft 8in)
Height: 2.24m (7ft 4in)
Engine: 123kW (165hp) GMC M10 diesel
Speed: 24km/h (15mph)
Range: 145km (90 miles)
Armament: 1 x 76mm (3in) 17pdr OQF, plus 1 x 7.7mm (0.303in) Bren MG
Radio: Wireless Set No. 18 CSL/126/14

Fortunately for the Allies, the German response to the landings was fatally slowed by Rommel's absence on leave and the need to get Hitler's permission to move the reserves. As a result only 21st Panzer Division was close enough to counterattack on D-Day – by 20:00 several of its AFVs with a *Panzergrenadier* company had managed to advance as far as the coast between Sword and Juno Beaches. Although it lacked the strength to do anything more than impose a temporary delay on the British and Canadian advance, the division showed what might have been possible if Rommel had been allowed to station all the available armour in readiness to launch immediate counterattacks whilst the Allies were at their most vulnerable.

The Allied invasion plans had assumed that Caen and Bayeux would be taken on D-Day itself, with all the beaches linked except for Utah, and that the front line would be 10–16km (6–10 miles) inland. Determined German resistance held the Allied advance well short of most of its key objectives, but the casualties had not been as heavy as some had feared (around 10,000 compared with the 20,000 Churchill had estimated) and the bridgeheads had withstood the expected counterattacks. The success or failure of the landings now hung on the outcome of the 'reinforcement race' – that is, whether Allied reinforcements could be brought across the Channel quickly enough to exceed the rate at which German forces could be moved in to seal off and destroy the beachheads.

Swings and roundabouts

Both sides faced severe problems – a prolonged spell of bad weather could wreck the prefabricated Mulberry Harbours, prevent resupply and ground Allied aircraft, giving the Germans a superb opportunity to attack unhindered by Allied air

▲ **Reconnaissance**
The crew of a US M8 armoured car observe a bombed-out building in Normandy, July 1944.

D-DAY TO ARNHEM: 1944

supremacy. Meanwhile, 'on the other side of the hill', the Germans had to move formations over a road and rail system under constant attack from Resistance groups and the Allied air forces. Furthermore, there was always the threat of further landings to consider – Allied deception measures included a very convincing fake First US Army Group ('FUSAG'), supposedly preparing to invade the Pas de Calais. As a result, German units that might have had a major impact on the fighting in Normandy were held back to face this phantom threat.

Allied reinforcement measures were quick and effective. Once the beachhead had been established, the components of two Mulberry Harbours were towed across the Channel. Both were assembled and operating by D+3 (9 June). The British Mulberry was set up at Arromanches, whilst its US counterpart was assembled at Omaha Beach. By 19 June, when severe storms delayed the landing of supplies for several days and destroyed the Omaha Mulberry, the British had

> **INSIGNIA**
>
> The badge of the French 2nd Armoured Division (*2e Division Blindée*) features the cross of Lorraine, emblem of the Free French, against a map of France.

▶ M3A3 Light Tank
US Third Army / V Corps / French 2nd Armoured Division (2e Division Blindée, 2e DB) / 1er Regiment de Marche de Spahis Marocains

General Leclerc's 2e DB scored its greatest triumph in the liberation of Paris on 24/25 August 1944. The M3s of the divisional reconnaissance unit were amongst the first French troops to enter the city.

Specifications

Crew: 4
Weight: 14.7 tonnes (14.46 tons)
Length: 5.02m (16ft 5in)
Width: 2.52m (8ft 3in)
Height: 2.57m (8ft 5in)
Engine: 186.25kW (250hp) Continental W-670-9A 7-cylinder radial petrol
Speed: 50km/h (31mph)
Range: 217km (135 miles)
Armament: 1 x 37mm (1.5in) M6 gun, plus 2 x 7.62mm (0.3in) MGs (1 coaxial, 1 ball-mounted in hull front)
Radio: SCR508

▲ M8 Armoured Car
French First Army / HQ 1st Armoured Division (1re Division Blindée)

In 1944–45 the French Army's armoured formations were closely based on the US model and the vast majority of their equipment was of US origin.

Specifications

Crew: 4
Weight: 8.12 tonnes (8 tons)
Length: 5m (16ft 5in)
Width: 2.54m (8ft 4in)
Height: 2.25m (7ft 5in)
Engine: 82kW (110hp) Hercules JXD 6-cylinder petrol
Speed: 89km/h (55mph)
Range: 563km (350 miles)
Armament: 1 x 37mm (1.5in) M6 gun, plus 1 x 12.7mm (0.5in) AA HMG and 1 x coaxial 7.62mm (0.3in) MG
Radio: SCR508

D-DAY TO ARNHEM: 1944

landed 314,547 men, 54,000 vehicles and 103,600 tonnes (102,000 tons) of supplies, and the Americans had put ashore 314,504 men, 41,000 vehicles and 117,850 tonnes (116,000 tons) of supplies. About 9100 tonnes (9000 tons) of stores were landed daily at the Arromanches Mulberry until the end of August, by which time the port of Cherbourg (captured on 27 June) was again fully operational.

In the vital first weeks of the Normandy campaign, the Germans were able to contain the Allied beachhead by matching their reinforcement rate. Their defence was greatly helped by the terrain – the infamous bocage with its thick, banked hedges and narrow lanes, dotted with villages that formed natural strongpoints. In this setting the German anti-tank guns and Panzerfausts were able to take a heavy toll

▲ M20 Armoured Utility Car
French First Army / HQ 1st Armoured Division (1re Division Blindée)

1re DB landed in Provence on 15 August 1944 as part of the Garbo Force under the command of General Jean de Lattre de Tassigny's French First Army. It took part in the liberation of Toulon and Marseilles and was the first Allied formation to reach the Rhone (25 August), the Rhine (19 November) and the Danube (21 April 1945).

Specifications

Crew: 4
Weight: 7 tonnes (6.89 tons) (estimated)
Length: 5m (16ft 5in)
Width: 2.54m (8ft 4in)
Height: 2m (6ft 7in) (estimated)
Engine: 82kW (110hp) Hercules JXD 6-cylinder petrol
Speed: 89km/h (55mph)
Range: 563km (350 miles)
Armament: 1 x 12.7mm (0.5in) AA HMG
Radio: SCR508

Specifications

Crew: 6
Weight: 25.9 tonnes (25.5 tons)
Length: 6.12m (20ft 1in)
Width: 2.71m (8ft 11in)
Height: 2.43m (8ft)
Engine: 298kW (400hp) Continental 9-cylinder radial petrol
Speed: 38km/h (24mph)
Range: 200km (125 miles)
Armament: 1 x 87.6mm (3.45in) 25pdr gun/howitzer, plus 2 x 7.7mm (0.303in) Bren MGs
Radio: Wireless Set No. 19

▲ Sexton SP 25pdr
Twenty-first Army Group / Second Army / XXX Corps / 8th Armoured Brigade / 147th Field Regiment RA

The Sexton was a highly successful attempt to produce a British equivalent of the Priest self-propelled howitzer. For whilst the Priest was a very good AFV, its 105mm (4.1in) howitzer was not a standard British weapon and the necessity for special supply arrangements caused constant problems. The first of over 2000 Sextons were issued in 1943 and many remained in service until 1956.

D-DAY TO ARNHEM: 1944

▲ Ram OP/Command Tank
Twenty-first Army Group / Second Army / XXX Corps / 8th Armoured Brigade / 147th Field Regiment RA

The Ram was a Canadian design based on the Sherman. Although it never saw combat service as a cruiser tank, 84 were converted to armoured observation posts for the Forward Observation Officers (FOOs) of Sexton self-propelled gun units. The main gun was replaced with a dummy 6pdr and two radios were fitted.

Specifications

Crew: 5
Weight: 29.48 tonnes (29 tons)
Length: 5.79m (19ft)
Width: 2.77m (9ft 1in)
Height: 2.67m (8ft 9in)
Engine: 298kW (400hp) Continental R-975 9-cylinder radial petrol
Speed: 40km/h (25mph)
Range: 232km (144 miles)
Armament: 1 x 7.62mm (0.3in) MG in sub-turret
Radio: 2 x Wireless Set No. 19

ORGANIZATION, JUNE 1944

- 7th Armoured Division HQ
  - 22nd Arm Bde HQ
    - 1 RTR HQ
    - 5 RTR HQ
    - 4th C Lon Yeo HQ
    - 1st Rfl Bde HQ
  - 131st Bde HQ
    - 1/5th Queens HQ
    - 1/6th Queens HQ
    - 1/7th Queens HQ
    - 2nd Devonshire HQ
    - 9th DLI HQ
    - Northumberland Fus HQ
  - Divisional Troops HQ
    - 8th Hussars HQ
    - 11th Hussars HQ
  - Royal Artillery HQ
    - 3rd RHA HQ
    - 5th RHA HQ
    - 15th Lt AA Rgt HQ
    - 65th A-T Rgt HQ

D-DAY TO ARNHEM: 1944

◀ **Pushing through the bocage**
A Churchill of 7 RTR, 31st Tank Brigade, moving up with infantry of 8th Royal Scots during Operation *Epsom*, 28 June 1944.

INSIGNIA

The 'Desert Rat' emblem was based on a design by the wife of 7th Armoured Ddivision's commander General Michael Creagh, giving rise to the legendary nickname.

▲ **Cromwell Mark IV**
7th Armoured Division / 22nd Armoured Brigade / 1 RTR

By the time that it went into action in Normandy, the Cromwell had fallen a generation behind the latest German tanks – although roughly equivalent to contemporary versions of the Panzer IV, it was far outclassed by the Panther. Despite its thin armour, the design had some good points, especially its high speed and reliability, which were fully exploited after the breakout from the Normandy beachhead.

Specifications
Crew: 5
Weight: 27.94 tonnes (27.5 tons)
Length: 6.35m (20ft 10in)
Width: 2.9m (9ft 6in)
Height: 2.49m (8ft 2in)
Engine: 447kW (600hp) Rolls-Royce Meteor V12 petrol
Speed: 64km/h (40mph)
Range: 280km (174 miles)
Armament: 1 x 75mm (2.9in) OQF gun, plus 2 x 7.92mm (0.31in) Besa MGs (1 coaxial and 1 ball-mounted in hull front)
Radio: Wireless Set No. 19

▲ **Infantry Tank Mark IV, Churchill Mark III**
Twenty-first Army Group / Second Army / VIII Corps / 15th (Scottish) Infantry Division / 31st Tank Brigade / 7 RTR

By mid-1944, the earlier Churchills were steadily being replaced by the Churchill Mark VII, which was armed with the Ordnance Quick Firing (OQF) 75mm (2.9in) gun, essentially the 6pdr bored out to fire US 75mm (2.9in) ammunition. Although the new gun had a greatly improved performance when firing HE in the tank's infantry support role, its armour-piercing capability was far less than that of the 6pdr. This situation often had lethal consequences when facing the newer, more heavily armoured German AFVs, and some units retained a number of 6pdr-armed tanks to give anti-tank protection.

Specifications
Crew: 5
Weight: 39.62 tonnes (39 tons)
Length: 7.44m (24ft 5in)
Width: 2.74m (9ft)
Height: 3.25m (10ft 8in)
Engine: 261.1kW (350hp) Bedford 12-cylinder petrol
Speed: 25km/h (15.5mph)
Range: 193km (120 miles)
Armament: 1 x 57mm (2.24in) 6pdr OQF gun, plus 2 x 7.92mm (0.31in) Besa MGs (1 coaxial and 1 ball-mounted in hull front)
Radio: Wireless Set No. 19

D-DAY TO ARNHEM: 1944

Specifications

Crew: 4
Weight: 32.7 tonnes (32.18 tons)
Length: 7.85m (25ft 9in)
Width: 2.67m (8ft 9in)
Height: 2.74m (8ft 11in)
Engine: 316.6kW (425hp) Chrysler Multibank A57 petrol
Speed: 40km/h (25mph)
Range: 161km (100 miles)
Armament: 1 x 76mm (3in) 17pdr OQF, plus 1 x coaxial 7.62mm (0.3in) MG
Radio: Wireless Set No. 19

▼ Sherman Firefly
Twenty-first Army Group / Second Army / I Corps / 3rd Canadian Infantry Division / 2nd Canadian Armoured Brigade / 1st Hussars

The Firefly was a brilliantly engineered conversion of the Sherman armed with the potent 17pdr gun. The first vehicles began to equip British and Canadian units shortly before D-Day, but were in short supply and initially had to be issued on the basis of one per troop, giving each troop one Firefly and three standard Shermans.

◀ Ambush
A Firefly lies in wait for a counterattack.

east of Caen with three armoured divisions, aimed at pinning down the Panzers and preventing their being transferred to counter the imminent American breakout from their western beachheads. Despite carpet-bombing (4500 Allied aircraft dropped 7100

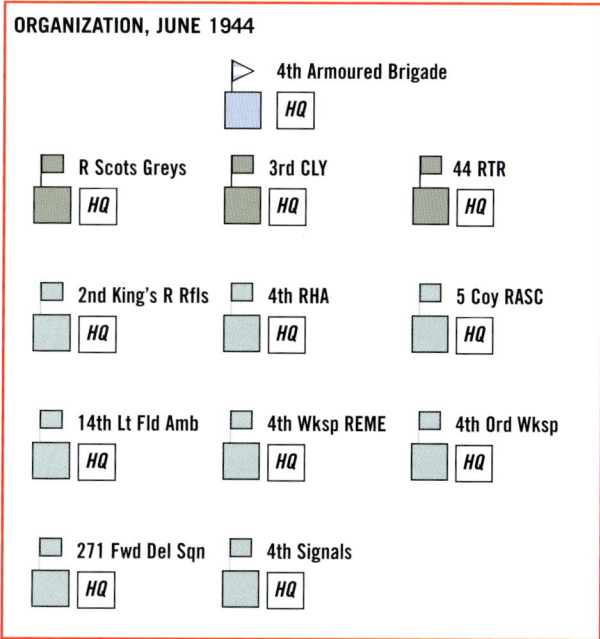

of Allied armour. When either side attempted to attack in these conditions, it was liable to suffer crippling losses. An example was the mauling of 7th Armoured Division's advance guard at Villers-Bocage by four Tiger Is and a single Panzer IV of *Schwere SS-Panzer Abteilung* 101.

Epsom and Charnwood

The first British and Canadian offensives, Operations *Epsom* and *Charnwood*, made little progress against a determined German defence, although the ruins of Caen were finally taken on 10 July. Then on the 18th, Montgomery launched Operation *Goodwood* south-

D-DAY TO ARNHEM: 1944

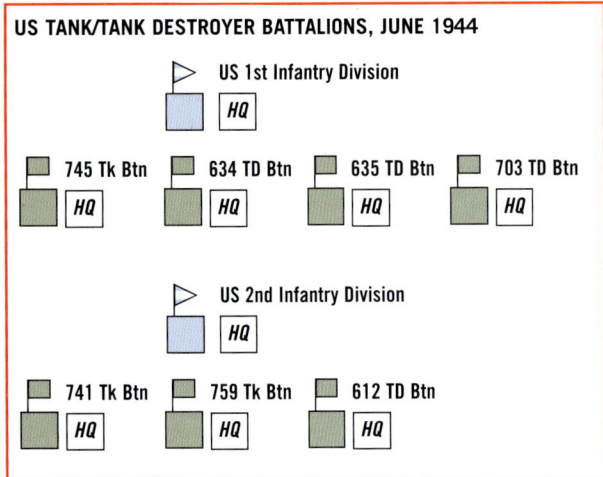

tonnes/7000 tons of bombs), three days of fierce fighting produced an advance of no more than 11km (7 miles) for the loss of almost 400 British and perhaps 60 German AFVs.

The US offensive, Operation *Cobra*, began on 25 July, and after 48 hours, broke through the weakened German lines, beginning one of the most dramatic advances of the war. By early August, Patton's armour was flooding south and east, threatening the rear of the German forces holding the line near Caen.

Four German divisions fielding 250 tanks counterattacked near Mortain on 7/8 August with the objective of cutting the American forces in two by driving through to the coast at Avranches. This desperate move was easily held and made the encirclement of the German Seventh Army and Fifth Panzer Army in the Falaise Pocket that much easier. When the pocket was finally overrun on 21 August, only 1300 Germans with 60 guns and 24 AFVs had escaped.

Limited supply

For a few weeks, it seemed as if the war in the West might be over – Paris was taken by the French 2nd Armoured Division (*2e Division Blindée, 2e DB*) on 25 August and Brussels was liberated by the Guards Armoured Division on 3 September. Then reality set in – on 4 September, supply shortages halted the Allied advance. Supply routes stretched back to the original invasion beaches and the nearby deep-water port of Cherbourg. Antwerp had been captured by Montgomery's Twenty-first Army Group, but its port facilities were unusable until German troops could be cleared from the approaches along the Scheldt estuary. Other Channel ports were either still under repair or, like Dunkirk, still in German hands. (The garrison of Dunkirk held out until May 1945.)

Frustratingly, although over-the-beach supply operations exceeded expectations and enough supplies were landed to support Allied operations,

▲ M10 Tank Destroyer
US Twelfth Army Group / First Army / VII Corps / 1st Infantry Division / 703rd Tank Destroyer Battalion

By the early stages of the Normandy campaign, the M10's 76mm (3in) gun was becoming less effective against the increasingly common up-armoured German AFVs. A further problem was the very slow traverse rate of its hand-cranked turret, which took about two minutes to rotate 360 degrees. These shortcomings led to the M10's replacement by the M36, with a much more potent 90mm (3.5in) gun.

Specifications

Crew: 5
Weight: 29.05 tonnes (28.6 tons)
Length: 5.82m (19ft 1in)
Width: 3.05m (10ft)
Height: 2.49m (8ft 2in)
Engine: 305.45kW (410hp) General Motors 6046 12-cylinder twin in-line diesel
Speed: 48km/h (30mph)
Range: 320km (200 miles)
Armament: 1 x 76mm (3in) M7 gun, plus 1 x 12.7mm (0.5in) HMG on AA mount
Radio: SCR610

these could not be taken forward due to transport shortages. At the beginning of September, 71,000 tonnes (70,000 tons) of supplies were stockpiled at Cherbourg because there were no means of moving them. Railway services were non-existent for months whilst damage from pre-invasion bombing was repaired – very limited rail movement out of Normandy only resumed on 30 August. A final blow to the hard-pressed supply services came when 1400 British 3-ton trucks were found to be useless because of faulty engine pistons — they could have moved 813 tonnes (800 tons) per day, enough to support two divisions.

These dire problems exacerbated disagreements amongst the senior Allied commanders – Montgomery and Patton's rivalry had remained fierce

Specifications

Crew: 7
Weight: 26.01 tonnes (25.6 tons)
Length: 6.02m (19ft 9in)
Width: 2.88m (9ft 5in)
Height: 2.54m (8ft 4in)
Engine: 298kW (400hp) Continental R975 C1 radial petrol
Speed: 42km/h (26mph)
Range: 201km (125 miles)
Armament: 1 x 105mm (4.1in) M1A2 howitzer, plus 1 x 12.7mm (0.5in) HMG on 'pulpit' AA mount.
Radio: SCR608

▲ **105mm Howitzer Motor Carriage (HMC) M7 (Priest)**
US First Army / V Corps / 2nd Armored Division / 14th Armored Field Artillery Battalion

Each US armoured division had three armoured field artillery battalions, each with 18 105mm (4.1in) M7s.

▲ **155mm Gun Motor Carriage (GMC) M12**
US Twelfth Army Group / First Army / VII Corps / 4th Infantry Division / 630th Heavy Artillery Battalion

The M12 was a World War I-vintage 155mm (6.1in) gun mounted on a heavily modified M3 Lee medium tank. Only 100 vehicles were completed before production ended in 1943 and most were used for training or put into storage. A total of 74 were refurbished and issued to heavy artillery battalions taking part in the Normandy landings. The type proved to be highly successful and remained in service until the end of the war.

Specifications

Crew: 6
Weight: 29.46 tonnes (29 tons)
Length: 6.73m (22ft 1in)
Width: 2.67m (8ft 9in)
Height: 2.69m (8ft 10in)
Engine: 263kW (353hp) Continental R-975 radial petrol
Speed: 39km/h (24mph)
Range: 225km (140 miles)
Armament: 1 x 155mm (6.1in) M1918M1 gun
Radio: SCR608

D-DAY TO ARNHEM: 1944

ever since the invasion of Sicily. Now both pressed claims that their forces alone could win the war with a single offensive deep into Germany, provided that they were given priority for the supplies that could be brought forward.

Operation *Market Garden*

Eventually it was agreed that Montgomery would be given just enough resources for his Operation *Market Garden*, a highly ambitious plan in which a major part of the First Allied Airborne Army would be dropped to seize the bridges at Eindhoven, Nijmegen and Arnhem. The formations to be inserted were the British 1st Airborne Division, the US 82nd and 101st Airborne Divisions and the 1st Polish Parachute Brigade. The intention was to form a so-called 'airborne carpet' more than 100km (63 miles) long across the Rhine over which the British Second Army would advance into Germany.

The operation, which began on 17 September, was hastily and poorly planned. Errors included the selection of drop zones too far from the bridges and a

◀ **Universal Carrier**

Twenty-first Army Group / Second Army / VIII Corps / 15th (Scottish) Infantry Division / 8th Royal Scots

The Universal Carrier was subjected to a vast range of modifications by its users – this example has simply been rearmed with a captured MG42 machine gun in the fighting compartment. A Bren gun, fitted with the rarely used 100-round drum magazine, is carried on the AA mount.

Specifications

Crew: 3
Weight: 4.06 tonnes (4 tons)
Length: 3.76m (12ft 4in)
Width: 2.11m (6ft 11in)
Height: 1.63m (5ft 4in)
Engine: 63.4kW (85hp) Ford V8 8-cylinder petrol
Speed: 52km/h (32mph)
Range: 258km (160 miles)
Armament: (Standard) 2 x 7.7mm (0.303in) Bren MGs

▲ **Tank Recovery Vehicle M32**

US First Army / V Corps / 2nd Armored Division / 66th Armored Regiment

The M32 was based on the chassis of the M4 medium tank with the turret replaced by a fixed superstructure. The vehicle was fitted with a 27,000kg (60,000lb) winch and a 5.5m (18ft) long pivoting A-frame jib. An 81mm (3.19in) mortar was mounted on the front of the hull and was primarily intended for firing white phosphorous smoke bombs to screen recovery work in front-line areas.

Specifications

Crew: 6
Weight: 29.2 tonnes (28.74 tons)
Length: 5.89m (19ft 4in)
Width: 2.62m (8ft 7in)
Height: 2.95m (9ft 8in)
Engine: 298kW (400hp) Continental R975 C1 9-cylinder radial petrol
Speed: 39km/h (24mph)
Range: 190km (120 miles)
Armament: 1 x 81mm (3.19in) mortar, plus 1 x 12.7mm (0.5in) AA HMG and 1 x 7.62mm (0.3in) MG ball-mounted in hull front

D-DAY TO ARNHEM: 1944

▲ **'Old Blood and Guts'**
Lieutenant-General George S. Patton was an undoubted master of 'hot pursuit' operations. One of the most notable of these was Operation *Cobra*, in which his Third Army advanced 97km (60 miles) in two weeks.

Stars displayed on a vehicle denoted its use by an officer of general rank. Patton's M20 carried this red panel with three stars to show that its occupant was a lieutenant-general.

▲ **M20 Armoured Utility Car**
US Twelfth Army Group / HQ Third Army
This modified M20 served as Patton's personal command vehicle.

Specifications

Crew: 4
Weight: 7 tonnes (6.89 tons) (estimated)
Length: 5m (16ft 5in)
Width: 2.54m (8ft 4in)
Height: 2m (6ft 7in) (estimated)

Engine: 82kW (110hp) Hercules JXD 6-cylinder petrol
Speed: 89km/h (55mph)
Range: 563km (350 miles)
Armament: 1 x 12.7mm (0.5in) AA HMG
Radio: SCR508

disregard for intelligence that correctly identified two SS Panzer divisions refitting in the Arnhem area. In addition, the Guards Armoured Division, leading the ground forces' advance, was committed to attack along a single road, Highway 69.

There was literally no room for manoeuvre, since much of the road, which rapidly became known as 'Hell's Highway', ran along causeways above polder terrain, which was too soft to support tactical vehicle movement. In the circumstances, it was hardly surprising that the results fell far short of the optimists' hopes of winning the war before Christmas. Guards Armoured Division took continuous losses from the fire of small groups of German AFVs as it advanced along the horribly exposed Highway 69, and had to fight off repeated

▲ **M3A1 Halftrack**

US First Army / V Corps / 2nd Armored Division / 41st Armored Infantry Regiment

Although its open-topped fighting compartment was vulnerable to air-burst artillery and grenades, the M3's high speed and mechanical reliability were invaluable during the rapid advances of Operation *Cobra*.

Specifications

Crew: 2 plus 11 passengers
Weight: 9.3 tonnes (9.15 tons)
Length: 6.34m (20ft 10in)
Width: 2.22m (7ft 3in)
Height: 2.69m (8ft 10in)
Engine: 109.5kW (147hp) White 160AX 6-cylinder petrol
Speed: 72km/h (45mph)
Range: 320km (200 miles)
Armament: 1 x 12.7mm (0.5in) HMG
Radio: SCR508

Specifications

Crew: 6
Weight: 7.97 tonnes (7.84 tons)
Length: 6.01m (19ft 9in)
Width: 1.96m (6ft 5in)
Height: 2.27m (7ft 5in)
Engine: 109.5kW (147hp) White 160AX 6-cylinder petrol
Speed: 72km/h (45mph)
Range: 320km (200 miles)
Armament: 1 x 81mm (3.19in) M1 mortar, plus 1 x 12.7mm (0.5in) HMG
Radio: SCR508

▲ **M4 Mortar Motor Carriage (MMC)**

US First Army / V Corps / 2nd Armored Division / 41st Armored Infantry Regiment

Each armoured infantry battalion had a mortar platoon with three M4s. These vehicles were later replaced by the M21 MMC, which mounted a forward-firing 81mm (3.19in) mortar.

D-DAY TO ARNHEM: 1944

counterattacks aimed at cutting the road. Although the division managed to link up with the US airborne force at Eindhoven and Nijmegen, it was held south of Arnhem by rapidly strengthening German forces. The remnants of the Allied airborne forces at Arnhem were evacuated on 25 September after holding out against overwhelming odds for nine days – more than twice as long as originally planned.

▲ Truck, 15cwt, GS, 4x2, Bedford MWD
Twenty-first Army Group / Second Army / XXX Corps / 7th Armoured Division / 22nd Armoured Brigade / 1st Rifle Brigade

The Bedford MWD was one of the best-known of the wartime 15cwts. It entered service just before the war, and Vauxhall Motors completed some 66,000 (not all of them GS trucks) by the time production ceased in 1945. In common with other 15cwts of the era, early models had an open cab and no tilt. They were initially issued on the basis of one per infantry platoon to carry the unit's ammunition, rations and kit. This example is finished in the so-called 'Mickey Mouse ear' camouflage widely used for British 'soft-skinned' support vehicles in 1944.

Specifications

Crew: 1
Weight: 2.13 tonnes (2.09 tons)
Length: 4.38m (14ft 4.5in)
Width: 1.99m (6ft 6.5in)
Height: 2.29m (7ft 6in)

Engine: 53.7kW (72hp) Bedford OHV 6-cylinder petrol
Speed: 80km/h (50mph) (estimated)
Range: 270km (168 miles) (estimated)

Specifications

Crew: 1
Weight: 6.56 tonnes (6.46 tons)
Length: 6.22m (20ft 5in)
Width: 2.18m (7ft 2in)
Height: 3.09m (10ft 2in)

Engine: 53.64kW (72hp) Bedford WD 6-cylinder petrol
Speed: 80km/h (50mph) (estimated)
Range: 450km (280 miles)

▲ Bedford 3-Ton Fuel Tanker
Twenty-first Army Group / Second Army / XXX Corps / 7th Armoured Division / 22nd Armoured Brigade / 5 RTR

Fuel was always a vital commodity and there never seemed to be enough tankers to meet demands. Conventional vehicles frequently had to make supplementary deliveries of thousands of jerricans of fuel to keep front-line units supplied.

Chapter 6

The Ardennes Offensive

In December 1944, it seemed as though the battered German forces were barely capable of effective defence. The accepted wisdom in Allied command circles was that the *Wehrmacht* would use the winter months to try to assemble some sort of strategic reserve to counter the offensives to be launched against the *Reich* in the spring of 1945. This was the logical military option, but most Allied commanders overlooked the fact that by this stage of the war German strategy was being shaped by Hitler's erratic intuition rather than by the highly professional officers of the German high command.

◀ **New kit**
US tank crews check their newly issued M24 Chaffee light tanks, December 1944.

THE ARDENNES OFFENSIVE

Wacht am Rhein
16 December 1944–24 January 1945

As early as August 1944, Hitler had directed that a force of at least 25 divisions should be made available for a major offensive in the West. By early December, this force had been assembled and was ready to be unleashed on a complacent and unsuspecting enemy.

IN THE AFTERMATH of Operation *Market Garden*, Allied forces closed up to the borders of the *Reich*, and the front from Switzerland to the North Sea seemed to settle into stalemate as winter closed in. This was partly due to increasingly effective resistance from German forces now holding shorter, more easily defensible lines. The other main factor was the seemingly interminable supply problem – which was alleviated but not solved by the opening of Antwerp in November 1944.

Hitler believed that this situation could be exploited by a major German offensive against one of the weak sectors of the long Allied front. Field-Marshals Walter Model and Gerd von Rundstedt favoured the so-called 'small solution' – a pincer attack near Aachen with the aim of trapping the US Third and Ninth Armies. Characteristically, Hitler rejected this plan in favour of a far more ambitious offensive through the Ardennes, with Antwerp as the

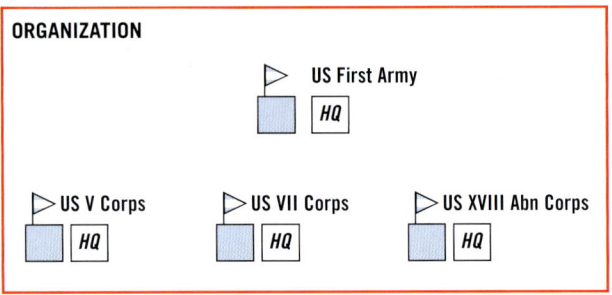

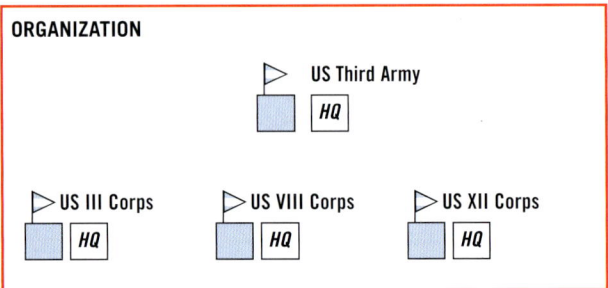

▲ **M4 Composite Medium Tank**
US Twelfth Army Group / Third Army / 6th Armored Division / 68th Armored Regiment

By 1944–45, it was rare to see such a pristine Sherman anywhere near the front-line. Almost all were festooned with spare track links, locally fitted appliqué armour plates and even sandbags in attempts to provide greater protection against the threat posed by Panzerfausts.

Specifications

Crew: 5
Weight: 31.8 tonnes (31.29 tons)
Length: 5.92m (19ft 5in)
Width: 2.62m (8ft 7in)
Height: 2.74m (8ft 11in)
Engine: 305.45kW (410hp) General Motors 6046 12-cylinder twin in-line diesel
Speed: 48km/h (30mph)
Range: 240km (150 miles)
Armament: 1 x 75mm (2.9in) M3 gun, plus 2 x 7.62mm (0.3in) MGs (1 coaxial, 1 ball-mounted in hull front)
Radio: SCR508

THE ARDENNES OFFENSIVE

objective. If successful, it would split the US and British forces, retake Brussels and Antwerp and trap four Allied armies. This victory in the West would have immense propaganda value, besides allowing the transfer of the bulk of the Panzer force to meet the Soviet advances in Poland and the Balkans.

The commanders who had to try to transform Hitler's vision into reality were privately scathing in their reactions to the plan. Wily old Field-Marshal von Rundstedt was appalled at the news that Antwerp was the objective. 'Antwerp?' he snorted. 'If we reach the Meuse we should go down on our knees and thank God!'

The raw and the weary

Hitler's choice of the Ardennes may have been influenced by his ambition to repeat the stunning victories of 1940. German security measures were highly effective, helped by the widespread Allied assumption that the *Wehrmacht* was simply too badly battered to be capable of launching any major offensive. A few Allied planners predicted the coming attack, including the chief intelligence officers of the US First and Third Armies, whose warnings were ignored by the US Twelfth Army Group, which had overall responsibility for the Ardennes front.

This sector of the front-line was thinly held by a high proportion of raw troops (notably the US 99th and 106th Infantry Divisions). These were supplemented by exhausted veteran formations such as the US 2nd Infantry Division, which had been sent to this 'quiet area' to rest and refit. Each division was

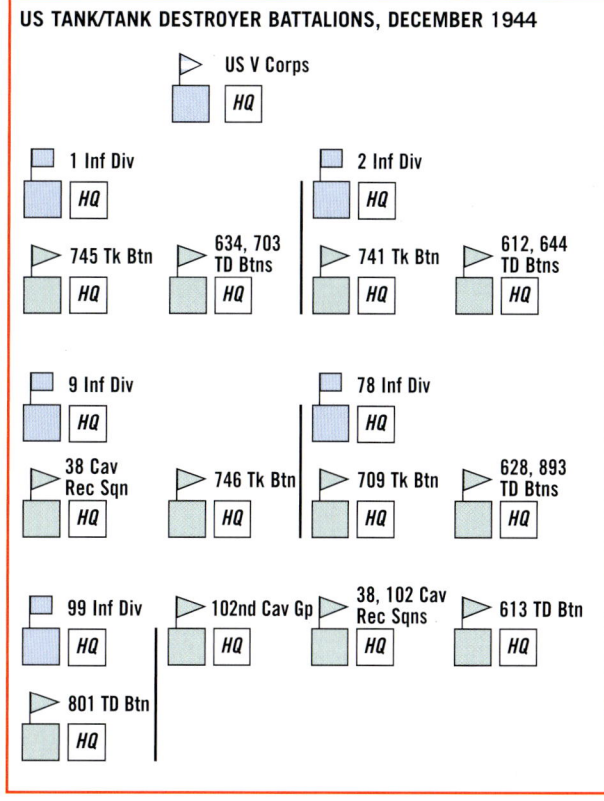

▲ **M4A3E8 Sherman Medium Tank**
US Twelfth Army Group / Third Army / 6th Armored Division / 68th Armored Regiment

The M4A3E8 'Easy Eight' introduced horizontal volute spring suspension (HVSS) and wider tracks, which considerably improved cross-country performance.

Specifications
Crew: 5
Weight: 33.7 tonnes (33.16 tons)
Length: 7.57m (24ft 10in)
Width: 3m (9ft 10in)
Height: 2.97m (9ft 9in)
Engine: 372.5kw (500hp) Ford GAA 8-cylinder petrol
Speed: 42km/h (26mph)
Range: 161km (100 miles)
Armament: 1 x 76mm (3in) M1A1 gun, plus 2 x 7.62mm (0.3in) MGs (1 coaxial and 1 ball-mounted in hull front)
Radio: SCR508

THE ARDENNES OFFENSIVE

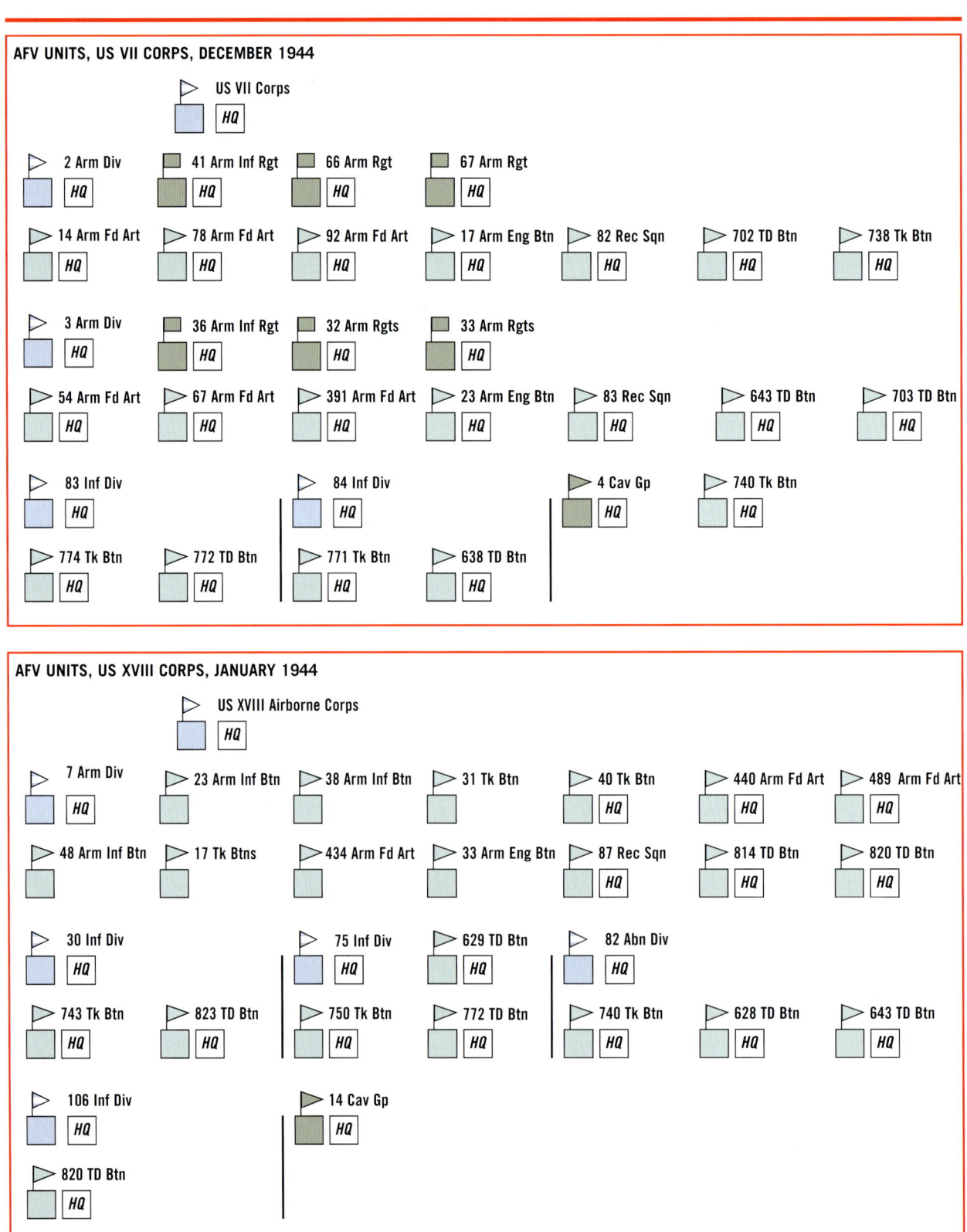

THE ARDENNES OFFENSIVE

badly overstretched, covering a front of roughly 45km (28 miles), three times the 'textbook figure'.

Offensive strength

The *Wehrmacht* and *Waffen*-SS forces assembling for the offensive were formidable but represented Germany's final strategic reserve. The main effort was to be made by Fifth Panzer Army, whose objective was Brussels, and Sixth SS Panzer Army, which was to take Antwerp. These two formations could field a total of 1500 AFVs, including some of the formidable Tiger II heavy tanks (which would prove to be dubious assets on the narrow, icy roads of the Ardennes).

The German forces also had to contend with serious problems – the success of the offensive was dependent on bad weather grounding the Allied air assets for long enough to allow the attacking forces to reach their objectives. This was risky enough, but dire fuel shortages meant that the Panzers would have to rely on captured stocks to cross the Meuse, let alone reach Brussels and Antwerp. In fact, the fuel crisis was so acute that the capture of Allied fuel dumps was added to the list of objectives.

The offensive began at 05:30 on 16 December 1944 with a massive artillery bombardment by 2000

▶ **Waiting for the panzers**
Sherman crews keeping watch in the Ardennes, late December 1944.

guns. By 08:00 three German armies were advancing through the Ardennes. In the northern sector SS General 'Sepp' Dietrich's Sixth SS Panzer Army assaulted the Losheim Gap and the Elsenborn Ridge in an attempt to break through to Liège. In the centre General Hasso von Manteuffel's Fifth Panzer Army attacked towards the key road junctions at Bastogne and St Vith. In the south General Erich

▲ **M24 Chaffee Light Tank**
US Twelfth Army Group / Third Army / 82nd Airborne Division / 740th Tank Battalion
Only two M24s were issued to the 740th Tank Battalion in time to see action during the Ardennes Offensive.

Specifications
Crew: 5
Weight: 18.28 tonnes (18 tons)
Length: 5.49m (18ft)
Width: 2.95m (9ft 8in)
Height: 2.46m (8ft 1in)
Engine: 2 x 82kW (110hp) Cadillac 44T24 V8 8-cylinder petrol
Speed: 55km/h (34mph)
Range: 282km (175 miles)
Armament: 1 x 75mm (2.9in) M6 gun, plus 1 x 12.7mm (0.5in) HMG on AA mount and 2 x 7.62mm (0.3in) MGs (1 coaxial, 1 ball-mounted in hull front)
Radio: SCR508

THE ARDENNES OFFENSIVE

Brandenberger's Seventh Army began a limited offensive towards Luxembourg to protect the left flank of the advance from Allied counterattacks.

The attacks by Sixth SS Panzer Army's infantry bogged down in the face of unexpectedly fierce resistance by the US 2nd and 99th Infantry Divisions on the Elsenborn Ridge, which caused Dietrich to commit his Panzer forces earlier than planned in an

AFV UNITS, US III CORPS, DECEMBER 1944

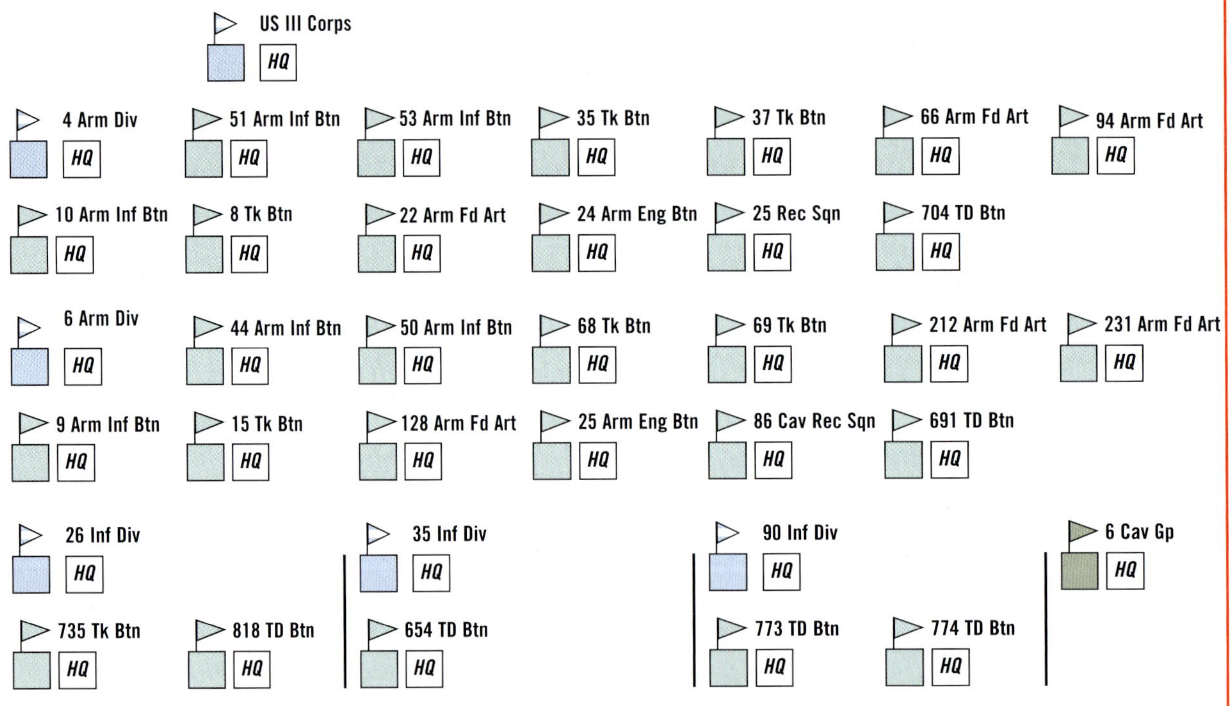

- US III Corps HQ
  - 4 Arm Div HQ
    - 51 Arm Inf Btn HQ
    - 53 Arm Inf Btn HQ
    - 35 Tk Btn HQ
    - 37 Tk Btn HQ
    - 66 Arm Fd Art HQ
    - 94 Arm Fd Art HQ
    - 10 Arm Inf Btn HQ
    - 8 Tk Btn HQ
    - 22 Arm Fd Art HQ
    - 24 Arm Eng Btn HQ
    - 25 Rec Sqn HQ
    - 704 TD Btn HQ
  - 6 Arm Div HQ
    - 44 Arm Inf Btn HQ
    - 50 Arm Inf Btn HQ
    - 68 Tk Btn HQ
    - 69 Tk Btn HQ
    - 212 Arm Fd Art HQ
    - 231 Arm Fd Art HQ
    - 9 Arm Inf Btn HQ
    - 15 Tk Btn HQ
    - 128 Arm Fd Art HQ
    - 25 Arm Eng Btn HQ
    - 86 Cav Rec Sqn HQ
    - 691 TD Btn HQ
  - 26 Inf Div HQ
    - 735 Tk Btn HQ
    - 818 TD Btn HQ
  - 35 Inf Div HQ
    - 654 TD Btn HQ
  - 90 Inf Div HQ
    - 773 TD Btn HQ
    - 774 TD Btn HQ
  - 6 Cav Gp HQ

▲ **105mm Howitzer Motor Carriage (HMC) M7 (Priest)**
US Twelfth Army Group / First Army / V Corps / 2nd Armored Division / 78th Armored Field Artillery Battalion

The M7 was the primary US self-propelled 105mm (4.1in) howitzer throughout the war. This example carries two pairs of logs for use as 'unditching beams'.

Specifications
Crew: 7
Weight: 26.01 tonnes (25.6 tons)
Length: 6.02m (19ft 9in)
Width: 2.88m (9ft 5in)
Height: 2.54m (8ft 4in)
Engine: 298kW (400hp) Continental R975 C1
Speed: 42km/h (26mph)
Range: 201km (125 miles)
Armament: 1 x 105mm (4.1in) M1A2 howitzer, plus 1 x 12.7mm (0.5in) HMG on 'pulpit' AA mount
Radio: SCR608

THE ARDENNES OFFENSIVE

attempt to achieve a breakthrough. This move was hampered by snow storms that rapidly created treacherous road conditions and inefficient traffic control that temporarily blocked several routes. In the centre Fifth Panzer Army advanced on a 30km (19-mile) front, forcing the surrender of two regiments (more than 8000 men) of the US 106th Division. The advance soon threatened St Vith and

AFV UNITS, US VIII CORPS, DECEMBER 1944

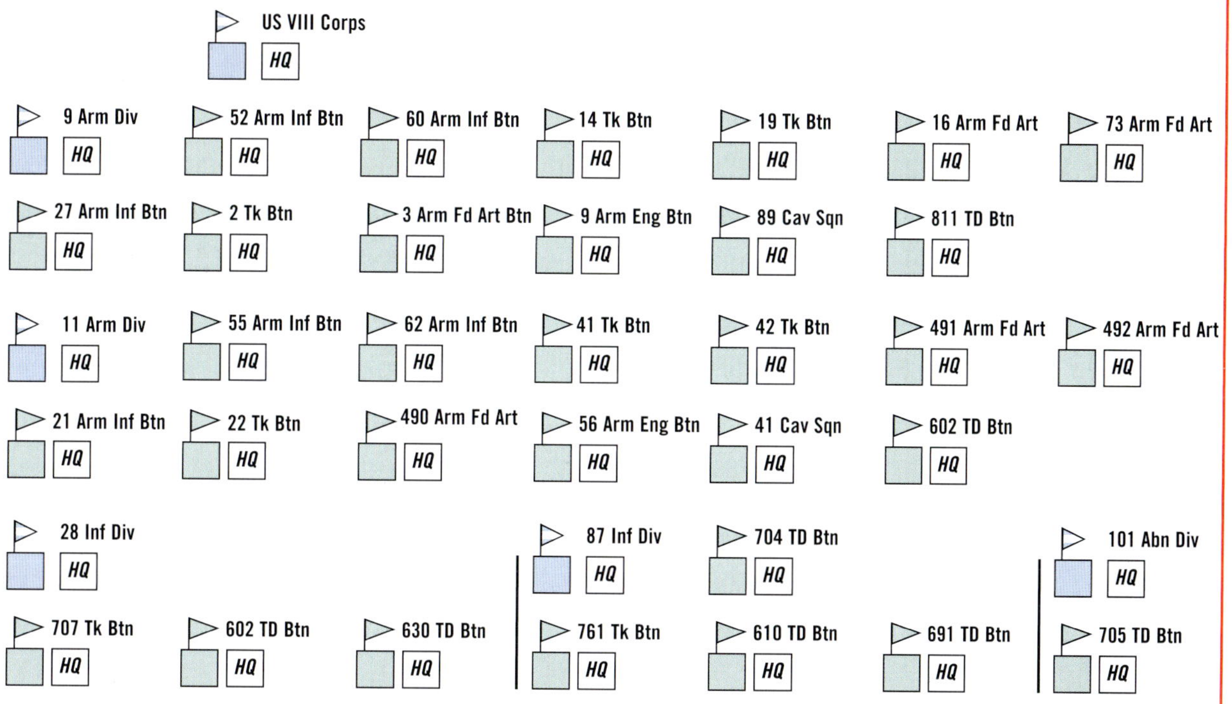

US VIII Corps HQ						
9 Arm Div HQ	52 Arm Inf Btn HQ	60 Arm Inf Btn HQ	14 Tk Btn HQ	19 Tk Btn HQ	16 Arm Fd Art HQ	73 Arm Fd Art HQ
27 Arm Inf Btn HQ	2 Tk Btn HQ	3 Arm Fd Art Btn HQ	9 Arm Eng Btn HQ	89 Cav Sqn HQ	811 TD Btn HQ	
11 Arm Div HQ	55 Arm Inf Btn HQ	62 Arm Inf Btn HQ	41 Tk Btn HQ	42 Tk Btn HQ	491 Arm Fd Art HQ	492 Arm Fd Art HQ
21 Arm Inf Btn HQ	22 Tk Btn HQ	490 Arm Fd Art HQ	56 Arm Eng Btn HQ	41 Cav Sqn HQ	602 TD Btn HQ	
28 Inf Div HQ			87 Inf Div HQ	704 TD Btn HQ		101 Abn Div HQ
707 Tk Btn HQ	602 TD Btn HQ	630 TD Btn HQ	761 Tk Btn HQ	610 TD Btn HQ	691 TD Btn HQ	705 TD Btn HQ

▲ M3 Halftrack
US Twelfth Army Group / Third Army / III Corps / 4th Armored Division / 53rd Armored Infantry Battalion

With only a canvas tilt for protection against the elements, all variants of the M3 were cold vehicles to operate in the depths of winter.

Specifications

Crew: 2 plus 11 passengers
Weight: 9.3 tonnes (9.15 tons)
Length: 6.34m (20ft 10in)
Width: 2.22m (7ft 3in)
Height: 2.69m (8ft 10in)

Engine: 109.5kW (147hp) White 160AX 6-cylinder petrol
Speed: 72km/h (45mph)
Range: 320km (200 miles)
Armament: 1 x 12.7mm (0.5in) HMG

THE ARDENNES OFFENSIVE

AFV UNITS, US XII CORPS, DECEMBER 1944

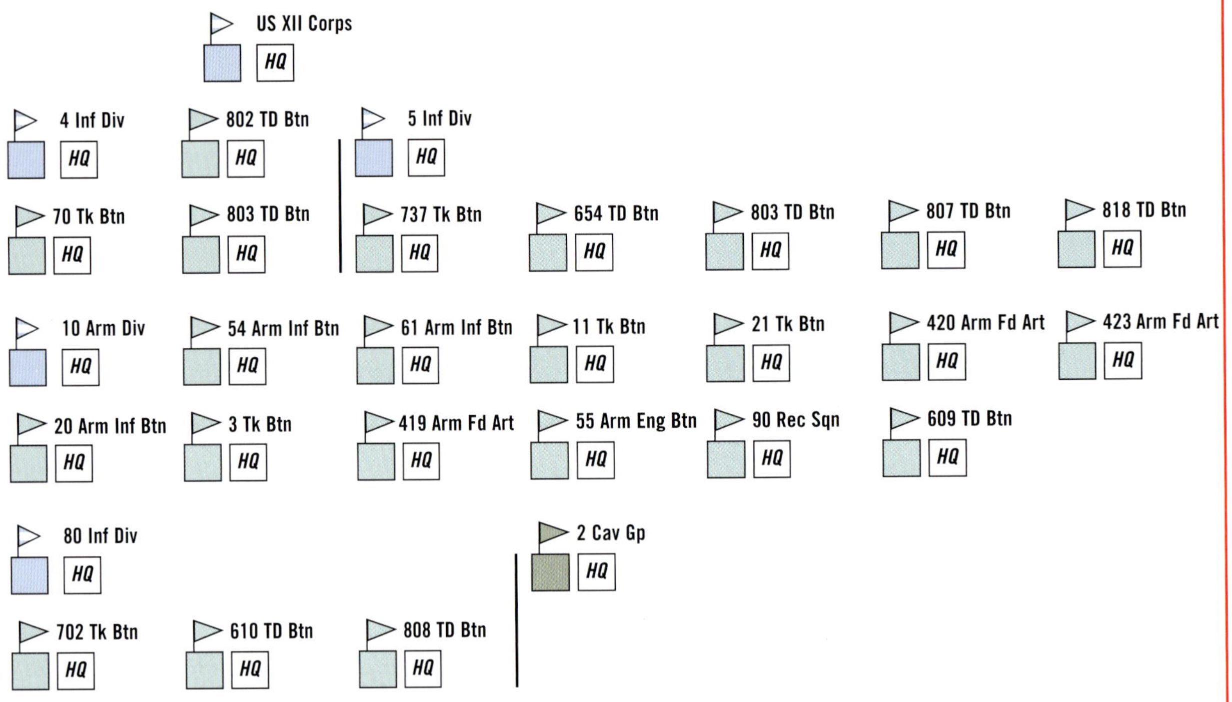

- US XII Corps HQ
  - 4 Inf Div HQ
    - 70 Tk Btn HQ
    - 10 Arm Div HQ
    - 20 Arm Inf Btn HQ
    - 80 Inf Div HQ
    - 702 Tk Btn HQ
  - 802 TD Btn HQ
    - 803 TD Btn HQ
    - 54 Arm Inf Btn HQ
    - 3 Tk Btn HQ
    - 610 TD Btn HQ
  - 5 Inf Div HQ
    - 737 Tk Btn HQ
    - 61 Arm Inf Btn HQ
    - 419 Arm Fd Art HQ
    - 808 TD Btn HQ
    - 654 TD Btn HQ
    - 11 Tk Btn HQ
    - 55 Arm Eng Btn HQ
    - 803 TD Btn HQ
    - 21 Tk Btn HQ
    - 90 Rec Sqn HQ
    - 2 Cav Gp HQ
    - 807 TD Btn HQ
    - 420 Arm Fd Art HQ
    - 609 TD Btn HQ
    - 818 TD Btn HQ
    - 423 Arm Fd Art HQ

Bastogne, whose defenders were reinforced by the US 82nd and 101st Airborne Divisions. German planners had counted on capturing St Vith by 17 December, but the town was only taken after its defenders withdrew on the 21st, having badly disrupted the German advance.

The 101st Airborne Division and Combat Command B of 10th Armored Division drove into Bastogne only hours before the town was encircled on 20 December. Four artillery battalions and the 705th Tank Destroyer Battalion also managed to reinforce the garrison, which was soon under attack by two Panzer and two infantry divisions from Fifth Panzer Army. 'Visualize the hole in a doughnut,' the 101st radioed Supreme Headquarters Allied Expeditionary Force (SHAEF) in Paris. 'That's us.' Bad weather grounded the Allied air forces, preventing air resupply or close air support, but the garrison held out, further slowing Fifth Panzer Army's advance by forcing it onto narrow minor roads. Despite this, its spearhead, 2nd Panzer Division, made the deepest penetration of the offensive, pushing to within 16km (10 miles) of the River Meuse on 24 December.

▶ **M3A1 Scout Car**

US Twelfth Army Group / Third Army / III Corps / HQ 4th Armored Division

Almost 21,000 M3A1s were produced between 1941 and 1944. Although partially superseded by the M8 and M20 armoured cars, large numbers remained in service throughout the war.

Specifications

Crew: 2 plus 6 passengers
Weight: 5.61 tonnes (5.53 tons)
Length: 5.62m (18ft 5in)
Width: 2.03m (6ft 8in)
Height: 2m (6ft 6in)
Engine: 71kw (95hp) White Hercules JXD 6-cylinder petrol
Speed: 105km/h (65mph)
Range: 400km (250 miles)
Armament: Up to 1 x 12.7mm (0.5in) HMG and 2 x 7.62mm (0.3in) MGs
Radio: SCR608

THE ARDENNES OFFENSIVE

▼ 703rd Tank Destroyer Battalion

By late 1944, the best-equipped tank destroyer battalions, such as the 703rd, were exceptionally effective anti-tank units. Their 90mm (3.5in) armed M36s had greater firepower than any other US AFVs until the first M26 Pershing heavy tanks began to reach front-line units in February 1945.

Battalion HQ (3 x M8 AC)

Reconnaissance Company

1 Platoon (3 x Jeeps, 1 x M8 AC)

2 Platoon (3 x Jeeps, 1 x M8 AC)

3 Platoon (3 x Jeeps, 1 x M8 AC)

Company x 3
1 Platoon (1 x Jeep, 2 x M8 ACs, 4 x M36 tank destroyers)

2 Platoon (1 x Jeep, 2 x M8 ACs, 4 x M36 tank destroyers)

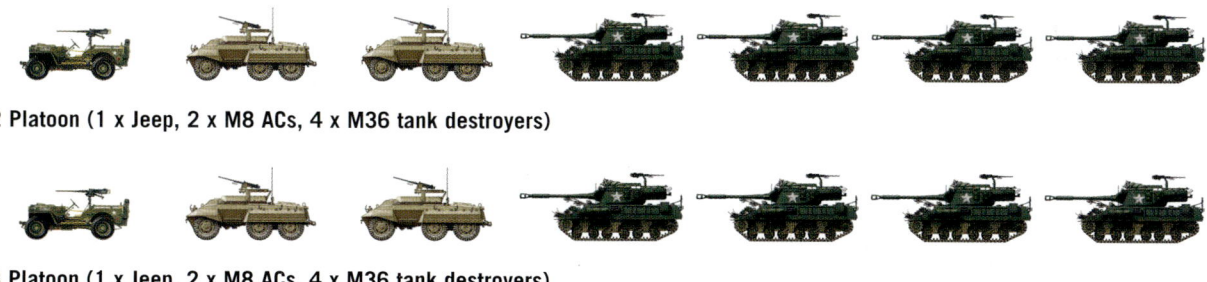

3 Platoon (1 x Jeep, 2 x M8 ACs, 4 x M36 tank destroyers)

Although good German security measures had ensured that the attacks achieved initial surprise, tenacious Allied defence of key points such as St Vith and Bastogne fatally slowed the tempo of the offensive. Whatever parallels Hitler may have drawn with the stunning victories won over the same ground in 1940 were misleading.

Conditions in 1944 were very different – the road network of the Ardennes had favoured the south-westerly axis of advance of the 1940 offensive, but *Wacht am Rhein* ('Watch on the Rhine', the codename for the 1944 operation) involved a north-westerly advance 'across the grain of the country'. Crucially, the Allied commanders of 1944 were very

THE ARDENNES OFFENSIVE

▲ **Ambush**
The commander of a well-dug-in M36 scans the horizon for targets.

different from those of 1940. Countermeasures were rapidly implemented as soon as it became clear that the operation was an all-out German offensive, rather than a limited spoiling attack. The US First Army and elements of Montgomery's Twenty-first Army Group to the north and west of the rapidly expanding German salient – the famous 'Bulge' – moved to protect the Meuse crossings. Away to the south, Patton swung the bulk of his US Third Army north to head for Bastogne. His progress was greatly aided by the failure of Brandenberger's Seventh Army to keep pace with the German advance to the north. This failure exposed Fifth Panzer Army's left flank to Patton's counterattack.

By 23 December, improving weather conditions allowed Allied air power to be brought into play

▲ **M3 Halftrack**
US Twelfth Army Group / Third Army / III Corps / 4th Armored Division / Medical Battalion
The versatile M3 was frequently employed as an armoured ambulance.

Specifications

Crew: 2 plus 11 passengers
Weight: 9.3 tonnes (9.15 tons)
Length: 6.34m (20ft 10in)
Width: 2.22m (7ft 3in)
Height: 2.69m (8ft 10in)

Engine: 109.5kW (147hp) White 160AX 6-cylinder petrol
Speed: 72km/h (45mph)
Range: 320km (200 miles)

◀ **M4 High-Speed Tractor**
US Twelfth Army Group / Third Army / VIII Corps / 87th Infantry Division / 912th Field Artillery Battalion
The M4 was based on the automotive components of the M2 light tank and entered service in 1942. It was widely used for towing heavy AA guns and 155mm (6.1in) guns and howitzers. A total of 5500 were produced from 1942 to 1945.

Specifications

Crew: 1 plus 11 passengers
Weight: 14.28 tonnes (14.06 tons)
Length: 5.23m (17ft 2in)
Width: 2.46m (8ft 1in)
Height: 2.51m (8ft 3in)

Engine: 156kW (210hp) Waukesha 145GZ 6-cylinder in-line petrol
Speed: 53km/h (33mph)
Range: 290km (180 miles)

THE ARDENNES OFFENSIVE

◀ M5 High-Speed Tractor
US Twelfth Army Group / Third Army / VIII Corps / 87th Infantry Division / 335th Field Artillery Battalion

The M5 was based on the tracks and suspension of the M3 light tank and was primarily used as a prime mover for 155mm (6.1in) howitzers.

Specifications
Crew: 1
Weight: 13.8 tonnes (13.58 tons)
Length: 5.03m (16ft 6in)
Width: 2.54m (8ft 4in)
Height: 2.69m (8ft 10in)

Engine: 154kW (207hp) Continental R6572
6-cylinder petrol
Speed: 48km/h (30 mph)
Range: 290km (180 miles)

▶ Dodge WC51 Weapons Carrier
US Twelfth Army Group / Third Army / III Corps / 4th Armored Division / 126th Ordnance Maintenance Battalion

The WC51 had a very good cross-country performance and could carry three times the load of a jeep. It was often referred to as a 'beep' – 'big jeep'.

Specifications
Crew: 1
Weight: 3.3 tonnes (3.24 tons)
Length: 4.47m (14ft 8in)
Width: 2.1m (6ft 10in)
Height: 2.15m (7ft)

Engine: 68.54kW (92hp) Dodge T214
6-cylinder petrol
Speed: 89km/h (55mph)
Range: 384km (240 miles)

▲ Studebaker US6 2½-Ton Truck
US Twelfth Army Group / Third Army / III Corps / 4th Armored Division / 126th Ordnance Maintenance Battalion

The Studebaker and GMC 2½-ton trucks played an essential role in supplying US armoured formations.

Specifications
Crew: 1
Weight: (loaded) 5.4 tonnes (5.3 tons)
Length: 6.82m (22ft 4.5in)
Width: 2.44m (8ft)
Height: 3.01m (9ft 10.5in)

Engine: 79kW (106hp) Hercules RXC
6-cylinder petrol
Speed: 64km/h (40mph)
Range: 255km (165 miles)

THE ARDENNES OFFENSIVE

against the German advance, which was already faltering as fuel began to run out. (Some fuel had been seized, but the major US fuel dumps had either been destroyed before capture or were successfully defended.) As ever, air attacks against AFVs only had limited success, the real damage being inflicted on their 'soft-skinned' support vehicles, without which the offensive was doomed. The effects of Allied air attacks were noted by Major-General Friedrich von Mellenthin, on his way to join 9th Panzer Division near Houffalize in the centre of the salient: 'The icebound roads glittered in the sunshine and I witnessed the uninterrupted air attacks on our traffic routes and supply dumps. Not a single German plane

Specifications
Crew: 5
Weight: 28.14 tonnes (27.7 tons)
Length: 6.15m (20ft 2in)
Width: 3.05m (10ft)
Height: 2.72m (8ft 11in)
Engine: 373kW (500hp) Ford GAA V8 petrol
Speed: 48km/h (30mph)
Range: 241km (150 miles)
Armament: 1 x 90mm (3.5in) M3 gun, plus 1 x 12.7mm (0.5in) AA HMG
Radio: SCR610

▲ **90mm Gun Motor Carriage (GMC) M36**
US Twelfth Army Group / Third Army / VIII Corps / 82nd Airborne Division / 703rd Tank Destroyer Battalion

The M36 was rushed into service with tank destroyer battalions in late 1944. It was highly popular with its crews as its 90mm (3.5in) gun could destroy almost any German AFV at normal battlefield ranges.

Specifications
Crew: 5
Weight: 18.18 tonnes (17.9 tons)
Length: 6.66m (21ft 10in)
Width: 2.97m (9ft 9in)
Height: 2.57m (8ft 5in)
Engine: 298.5kW (400hp) Continental R-975 9-cylinder radial petrol
Speed: 89km/h (55mph)
Range: 241km (150 miles)
Armament: 1 x 76mm (3in) M1 gun, plus 1 x 12.7mm (0.5in) AA HMG
Radio: SCR610

▲ **76mm Gun Motor Carriage (GMC) M18 'Hellcat'**
US Twelfth Army Group / Third Army / VIII Corps / 101st Airborne Division / 705th Tank Destroyer Battalion

The 705th played an important role in the defence of Bastogne, destroying 43 German AFVs for the loss of six Hellcats.

THE ARDENNES OFFENSIVE

was in the air and innumerable vehicles were shot up and their blackened wrecks littered the roads.'

Equally important, the improving weather allowed air resupply drops to the hard-pressed defenders of Bastogne from the morning of 23 December, after which a total of 1446 containers were parachuted into the perimeter. These containers were a crucial factor in the defence of the town, as artillery

▲ **Careful!**
A Sherman and infantry cautiously advance along a seemingly deserted street.

ammunition was almost exhausted – in many cases shells had to be rushed straight from the drop zones to the batteries.

Although much emphasis is rightly given to the importance of Allied air power, the battered *Luftwaffe*

▲ **Sherman Firefly**

Twenty-first Army Group / Second Army / XXX Corps / 29th Armoured Brigade / 2nd Fife and Forfar Yeomanry

The 29th Armoured Brigade was rushed to Dinant to protect the Meuse crossings in late December 1944. By now British armoured units were well on the way to achieving a fifty-fifty mix of Fireflies and 75mm (2.9in) armed Shermans.

Specifications
Crew: 4
Weight: 32.7 tonnes (32.18 tons)
Length: 7.85m (25ft 9in)
Width: 2.67m (8ft 9in)
Height: 2.74m (8ft 11in)
Engine: 316.6kW (425hp) Chrysler Multibank A57 petrol
Speed: 40km/h (24.8mph)
Range: 161km (100 miles)
Armament: 1 x 76mm (3in) 17pdr OQF, plus 1 x coaxial 7.62mm (0.3in) MG
Radio: Wireless Set No. 19

THE ARDENNES OFFENSIVE

▲ **Every little helps …**
An M5 light tank, with unusual 'appliqué armour' of split logs on the glacis plate, moves up to the front.

did its best to support the German offensive. Night ground-attack units equipped with a mixture of Fw 190s and Ju 87s attacked Allied AFVs, troops and supply lines on an almost nightly basis throughout the second half of December. These units were highly trained for this extremely demanding role and their aircraft were fitted with sophisticated 'blind-flying' equipment that allowed them to operate in weather conditions that grounded most other aircraft. In a typical sortie on the evening of 16 December, 50-plus Ju 87s attacked US positions around Monschau on the northern flank of the offensive, dropping flares before bombing and strafing the area.

Whenever possible, conventional ground-attack units also went into action, with their Fw 190s using the new 'Panzerblitz' anti-tank rockets against US armour around St Vith and Bastogne. However, the *Luftwaffe*'s efforts had only a limited impact, being crippled by fuel shortages and the overwhelming Allied numerical superiority.

On Christmas Day, a final German assault on Bastogne was beaten off and Patton's forces broke through to raise the siege the next day. Although it would take almost another month to retake the entire Bulge, the relief of Bastogne marked the end of the real crisis. On 3 January 1945 the Allies went over to the offensive, and on the 16th units of the US First and Third Armies joined hands at Houffalize, eliminating the bulk of the salient. The offensive had inflicted 19,000 casualties on US Twelfth Army Group, and had taken 15,000 American prisoners. But the cost to the German Army had been 100,000 men killed or wounded and 800 AFVs destroyed – losses that could not be made up.

What went wrong?

In the wooded hills of the Ardennes, the Tigers and Panthers were unable to exploit their superior long-range firepower and were themselves vulnerable to anti-tank guns and bazooka teams lying in ambush. German attempts to manoeuvre were hampered by the terrain, which frequently limited advances to a single vehicle front. In these situations, a single well-positioned roadblock could halt an entire division.

Some reports indicate that a handful of 2nd Panzer Division's Panthers fitted with prototype infra-red (IR) searchlights and night sights were used for combat trials during the offensive. These may have

THE ARDENNES OFFENSIVE

been instrumental in defeating Task Force Harper (a tank battalion plus two armored infantry companies of the US 9th Armored Division) holding the crossroads at Fe'itsch on the approaches to Bastogne, where they were credited with the destruction of at least 10 Shermans in a single night attack. However, the handful of IR-equipped Panthers were too few to make a decisive impact on the outcome of the operation. Equally, the dire fuel shortages caused by Allied bombing meant that even when manoeuvres to by-pass Allied positions were possible, units could find themselves stranded with empty fuel tanks, often within a few kilometres of key objectives.

Sepp Dietrich, commanding Sixth SS Panzer Army, bluntly summed up the unrealistic thinking behind the planning of the offensive, remarking that 'All I had to do was to cross the river, capture Brussels, and then go on to take the port of Antwerp. The snow was waist deep and there wasn't room to deploy four tanks abreast, let alone six armoured divisions. It didn't get light until eight and it was dark again at four, and my tanks can't fight at night. And all this at Christmas time!'

Albert Speer, *Reichminister* for Armaments and War Production, wrote, 'The failure of the Ardennes Offensive meant that the war was over.'

▲ M3A3 Stuart V Light Tank

Twenty-first Army Group / Second Army / XXX Corps / 29th Armoured Brigade / 23rd Hussars

Stuarts remained popular reconnaissance vehicles, but their relatively ineffective 37mm (1.5in) gun led to a significant number being converted to 'Stuart Recce' vehicles, with turrets removed and machine-gun pintle mounts fitted.

Specifications

Crew: 4
Weight: 14.7 tonnes (14.46 tons)
Length: 5.02m (16ft 5in)
Width: 2.52m (8ft 3in)
Height: 2.57m (8ft 5in)
Engine: 186.25kW (250hp) Continental W-670-9A 7-cylinder radial petrol
Speed: 50km/h (31mph)
Range: 217km (135 miles)
Armament: 1 x 37mm (1.5in) M6 gun, plus 3 x 7.62mm (0.3in) MGs (1 AA, 1 coaxial, 1 ball-mounted in hull front)
Radio: Wireless Set No. 19

◀ Daimler Scout Car Mark II

Twenty-first Army Group / Second Army / XXX Corps / 29th Armoured Brigade / 23rd Hussars

This Daimler has been fitted with a 7.62mm (0.3in) Browning machine gun in place of the more usual Bren.

Specifications

Crew: 2
Weight: 3.22 tonnes (3.2 tons)
Length: 3.23m (10ft 5in)
Width: 1.72m (5ft 8in)
Height: 1.5m (4ft 11in)
Engine: 41kW (55hp) Daimler 6-cylinder petrol
Speed: 89km/h (55mph)
Range: 322km (200 miles)
Armament: 1 x 7.62mm (0.3in) MG
Radio: Wireless Set No. 19

Chapter 7

Invading the Reich

The Ardennes Offensive had fallen far short of its grandiose aims, but it had badly delayed and disrupted Allied preparations for the advance to the Rhine. As a result, the battles to reach the Rhine were fought in appalling conditions of alternating frost and thaw, which posed tremendous problems for armoured operations. Such natural hazards were compounded by the threat posed by a skilful and determined enemy. Although most panzer formations facing the Western Allies were now far below strength, even ad hoc groupings of three or four Panthers or Tigers could decimate an unwary British or US tank unit.

◀ **Pershings at rest**
US Army M26 Pershing heavy tanks are stopped beside a railway line in occupied Germany, April 1945. The first Pershings were issued in January of the same year and proved formidable in combat.

INVADING THE REICH

To the Rhine
JANUARY–MARCH 1945

By early 1945, victory was in sight, but the Ardennes Offensive had been a bloody reminder of the dangers of underestimating an enemy who showed no inclination to give in.

THE ALLIED OFFENSIVES of 1945 began with Operation *Blackcock*, which was launched by Second Army's XII Corps (7th Armoured Division, 43rd Wessex Division and 52nd Lowland Division) on 14 January. The objective was to clear the 'Roer Triangle', an area straddling the Dutch-German border, which included outlying defences of the Siegfried Line. Almost every farm and hamlet in the area had to be attacked, often with the support of detachments of Crocodile flamethrower tanks from 79th Armoured Division.

In a single action at Susteren, 1 RTR lost seven tanks to a combination of Panzerfaust teams and anti-tank guns. Total losses incurred in the operation were 43 tanks, 20 to enemy action and the remaining 23 to accidents and breakdowns. The area was finally cleared on 26 January and preparations began for the next offensive, Operation *Veritable*.

Operation *Veritable*

Veritable's objective was the Reichswald, a heavily forested area about 25km (16 miles) west of Nijmegen between the Waal and Maas Rivers. In the winter of 1944/45 the Germans opened the sluices

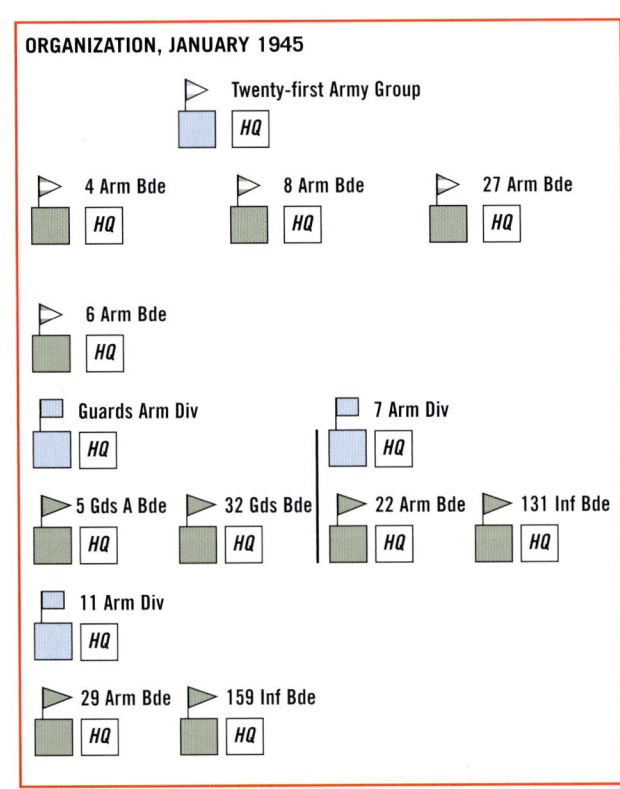

ORGANIZATION, JANUARY 1945

▸ **M22 Locust Light Tank**
Twenty-first Army Group / 6th Airborne Division / Airborne Armoured Reconnaissance Regiment
This was the American equivalent of the Tetrarch but was never used in action by US airborne forces as they had no suitable gliders. A few were used by 6th Airborne Division alongside their Tetrarchs in the Rhine crossings.

Specifications
Crew: 3
Weight: 7.41 tonnes (7.3 tons)
Length: 3.94m (12ft 11in)
Width: 2.24m (7ft 4in)
Height: 1.73m (5ft 8in)
Engine: 121kW (162hp) Lycoming O-435T 6-cylinder petrol
Speed: 64km/h (40mph)
Range: 217km (135 miles)
Armament: 1 x 37mm (1.5in) M6 gun, plus 1 x coaxial 7.62mm (0.3in) MG
Radio: SCR510

INVADING THE REICH

on these rivers so that the areas north and south of the forest were flooded and impassable, forcing the Allies into a frontal attack on the area, which included sections of the Siegfried Line defences.

The operation was controlled by Twenty-first Army Group, which deployed elements of the First Canadian, Second British and Ninth US Armies. A two-pronged attack was planned: the northern prong would consist of XXX Corps under the command of First Canadian Army, and the southern prong would consist of formations of Ninth US Army, striking across the River Roer shortly after XXX Corps attacked. In the initial stages of its attack, XXX Corps deployed the following units and formations: Guards Armoured Division; 15th, 43rd, 51st and 53rd Infantry Divisions plus 2nd and 3rd Canadian Infantry Divisions; 6th Guards Armoured Brigade plus 8th and 34th Armoured Brigades; 3rd, 4th, 5th, 9th and 2nd Canadian AGRAs (Army Groups Royal Artillery); as well as support units including two searchlight batteries, elements of 11 regiments of 79th Armoured Division and two Royal Engineers assault regiments.

The total strength of the corps was just over 200,000 all ranks. The offensive opened on 8 February, the attack preceded by an artillery barrage of over 6100 tonnes (6000 tons) of shells. Although roughly 500 Allied tanks were opposed by no more

▶ C15TA Armoured Truck

Twenty-first Army Group / Second Army / XII Corps / HQ 7th Armoured Division

The C15TA was developed by General Motors Canada based on the chassis of the Chevrolet C15 CMP truck. It was extensively used as an APC and armoured ambulance by British and Canadian units in 1944–45. A total of 3961 vehicles were produced in Canada between 1943 and 1945.

Specifications
Crew: 2 plus 8 passengers
Weight: 4.5 tonnes (4.42 tons)
Length: 4.75m (15ft 7in)
Width: 2.34m (7ft 8in)
Height: 2.31m (7ft 7in)
Engine: 74kW (100hp) GMC 6-cylinder petrol
Speed: 65km/h (40mph)
Range: 483km (300 miles)

▲ Sherman Firefly

Twenty-first Army Group / 7th Armoured Division / 22nd Armoured Brigade / 5th Royal Inniskilling Dragoon Guards

By early 1945 the 17pdr's anti-tank performance had significantly improved as APDS ammunition became more readily available.

Specifications
Crew: 4
Weight: 32.7 tonnes (32.18 tons)
Length: 7.85m (25ft 9in)
Width: 2.67m (8ft 9in)
Height: 2.74m (8ft 11in)
Engine: 316.6kW (425hp) Chrysler Multibank A57 petrol
Speed: 40km/h (24.8mph)
Range: 161km (100 miles)
Armament: 1 x 76mm (3in) 17pdr OQF, plus 1 x coaxial 7.62mm (0.3in) MG
Radio: Wireless Set No. 19

INVADING THE REICH

than 50 German tanks and 36 assault guns, the German armour included some Jagdpanthers plus six Sturmtigers armed with 380mm (14.9in) rocket launchers. Such opponents demanded respect – a single 350kg (770lb) rocket from a Sturmtiger was reported to have destroyed three US Shermans in January 1945.

The effectiveness of these small numbers of heavy German AFVs was enhanced by the terrain. Allied forces could not exploit their numerical superiority on the narrow front of less than 10km (6 miles), where dense forests and deep mud largely confined AFVs to roads and tracks. (Even the renowned cross-country ability of the Churchills was tested to the limit – several were lost after becoming hopelessly bogged down.) Besides the handful of powerful German AFVs, Allied armour also had to contend with repeated ambushes by infantry anti-tank teams armed with Panzershrecks and Panzerfausts, who exploited the ample cover to make killing shots from close range. Given the circumstances, it was understandable that it took until 21 February to clear the entire area.

The situation was not helped by the enforced delay in launching Operation *Grenade*, the southern prong of the offensive. As had been feared, the Germans had very efficiently sabotaged the Roer dams, flooding a large area and imposing a two-week delay on Ninth Army's attack, which finally seized crossings over the river on 23 February. The experience of combat in such a harsh environment prompted a special report by 9 RTR, which vividly summarizes the lessons of the battle (see below, Appendices).

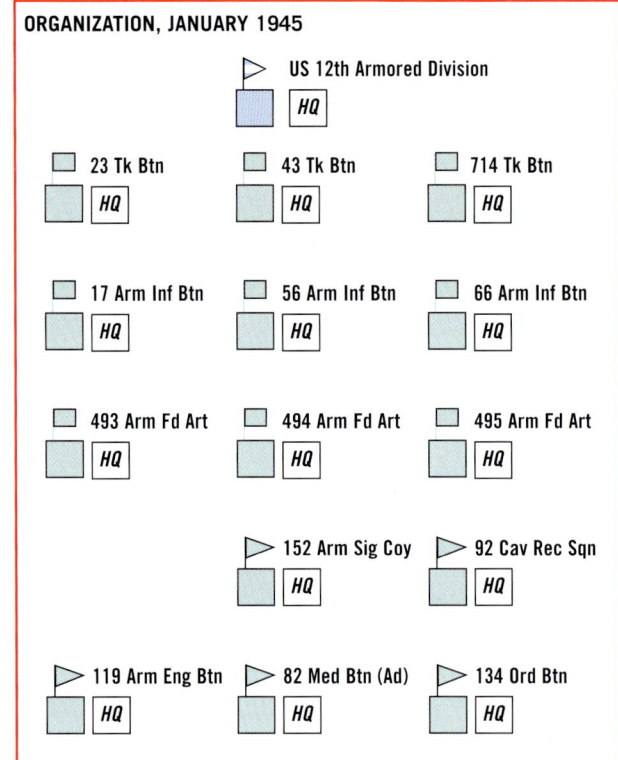

ORGANIZATION, JANUARY 1945

US 12th Armored Division / HQ

- 23 Tk Btn / HQ
- 43 Tk Btn / HQ
- 714 Tk Btn / HQ
- 17 Arm Inf Btn / HQ
- 56 Arm Inf Btn / HQ
- 66 Arm Inf Btn / HQ
- 493 Arm Fd Art / HQ
- 494 Arm Fd Art / HQ
- 495 Arm Fd Art / HQ
- 152 Arm Sig Coy / HQ
- 92 Cav Rec Sqn / HQ
- 119 Arm Eng Btn / HQ
- 82 Med Btn (Ad) / HQ
- 134 Ord Btn / HQ

▲ **Tractor, Artillery, 4x4, Morris C8 Mark III**
Twenty-first Army Group / Second Army / XXX Corps / Guards Armoured Division / 55th Field Regiment RHA
The C8 Mark III entered service in 1944 as a dual-purpose vehicle, capable of towing both the 17pdr anti-tank gun and the 25pdr gun/howitzer.

Specifications
Crew: 1 plus 7 passengers
Weight: 3.4 tonnes (3.34 tons)
Length: 4.49m (14ft 9in)
Width: 2.21m (7ft 3in)
Height: 2.26m (7ft 5in)
Engine: 52.2kW (70hp) Morris 4-cylinder petrol
Speed: 72km/h (45mph) (estimated)
Range: 322km (200 miles) (estimated)

INVADING THE REICH

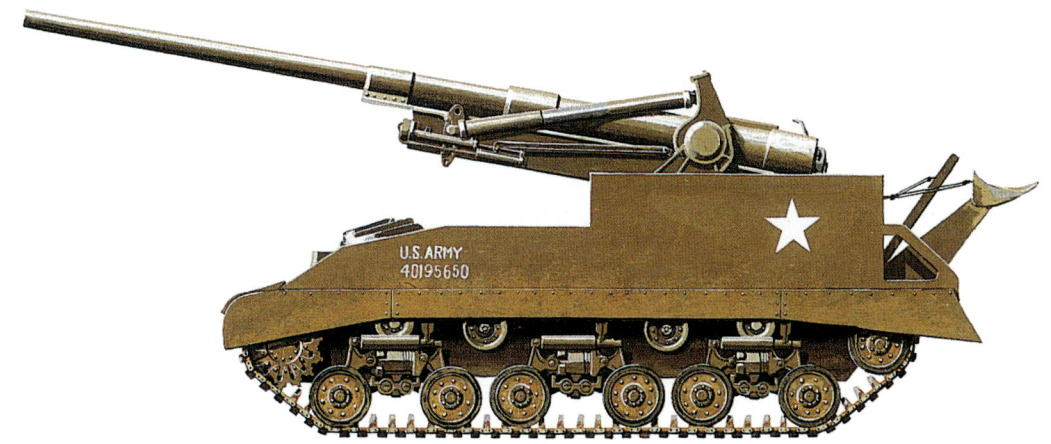

Specifications
Crew: 8
Weight: 40.64 tonnes (40 tons)
Length: 9.04m (29ft 9in)
Width: 3.15m (10ft 4in)
Height: 2.69m (8ft 10in)
Engine: 295kW (395hp) Continental 9-cylinder radial petrol
Speed: 39km/h (24mph)
Range: 161km (100 miles)
Armament: 1 x 155mm (6.1in) M1A1 gun
Radio: SCR608

▲ 155mm Gun Motor Carriage (GMC) M40
US Twelfth Army Group / First Army / 3rd Armored Division / 991st Field Artillery Battalion

The M40 comprised the 155mm (6.1in) M2 gun on a heavily modified hull of the M4A3 Sherman fitted with HVSS suspension. Only a single vehicle was issued for combat trials with 3rd Armored Division before the end of the war.

▲ M26 Pershing Heavy Tank
US Twelfth Army Group / Third Army / 9th Armored Division / 2nd Armored Regiment

A detachment of three Pershings attached to 9th Armored Division took part in the seizure of the Ludendorff Bridge across the Rhine at Remagen.

Specifications
Crew: 5
Weight: 41.86 tonnes (41.2 tons)
Length: 8.61m (28ft 3in)
Width: 3.51m (11ft 6in)
Height: 2.77m (9ft 1in)
Engine: 373kW (500hp) Ford GAF V8 petrol
Speed: 48km/h (30mph)
Range: 161km (100 miles)
Armament: 1 x 90mm (3.5in) M3 gun, plus 1 x 12.7mm (0.5in) AA HMG and 2 x 7.62mm (0.3in) MGs (1 coaxial and 1 ball-mounted in hull front)
Radio: SCR508/528

INVADING THE REICH

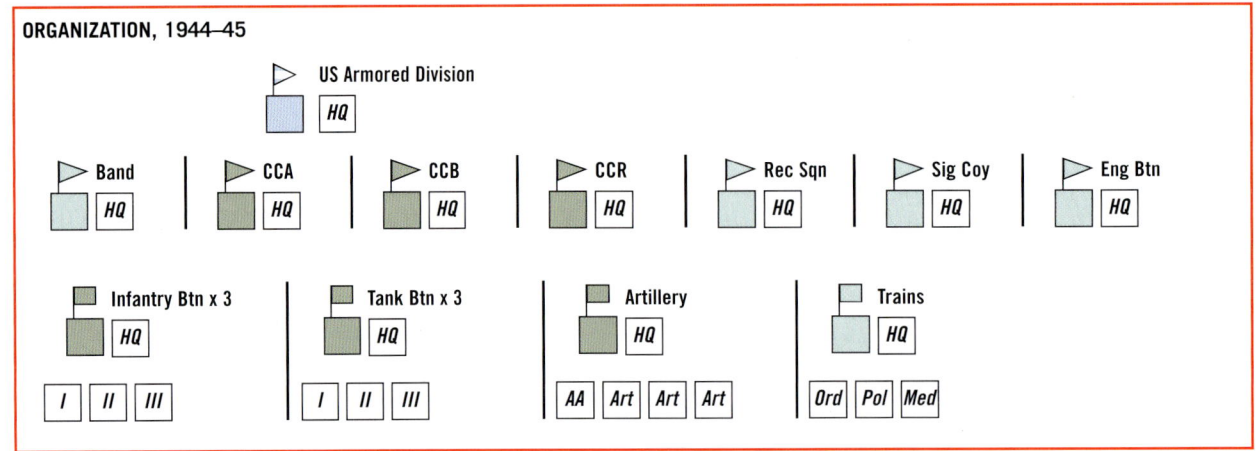

Principal Equipment, US Armored Division	Strength
Medium Tanks	186
Light Tanks	77
SP Howitzers 105mm (4.1in)	54
Carriers, Halftrack	501
MGs 7.62mm (0.3in)	465
MGs 12.7mm (0.5in)	404
Sub-machine Guns	2803
Carbines	5286
Rifles 7.62mm (0.3in)	2063

US Light Tank Company	Strength
Company HQ:	
¼-Ton Trucks	2
M5 Tanks	2
2½-Ton Trucks	1
1-Ton Trailers	1
M3 Halftracks	1
Light ARVs	1
Tank Platoons x 3:	
M5 Tanks	5

▲ **M24 Chaffee Light Tank**

US Twelfth Army Group / Third Army / 9th Armored Division / 14th Armored Regiment

The Chaffee was armed with a lightweight 75mm (2.9in) gun developed from a type used in the B-25H Mitchell bomber. The aircraft's gun was itself derived from the Sherman's M3 75mm (2.9in) and had the same ballistics, but used a thin-walled barrel and different recoil mechanism.

Specifications

Crew: 5
Weight: 18.28 tonnes (18 tons)
Length: 5.49m (18ft)
Width: 2.95m (9ft 8in)
Height: 2.46m (8ft 1in)
Engine: 2 x 82kW (110hp) Cadillac 44T24 V8 8-cylinder petrol
Speed: 55km/h (34mph)
Range: 282km (175 miles)
Armament: 1 x 75mm (2.9in) M6 gun, plus 1 x 12.7mm (0.5in) HMG on AA mount and 2 x 7.62mm (0.3in) MGs (1 coaxial, 1 ball-mounted in hull front)
Radio: SCR508

INVADING THE REICH

ORGANIZATION, 1944–45

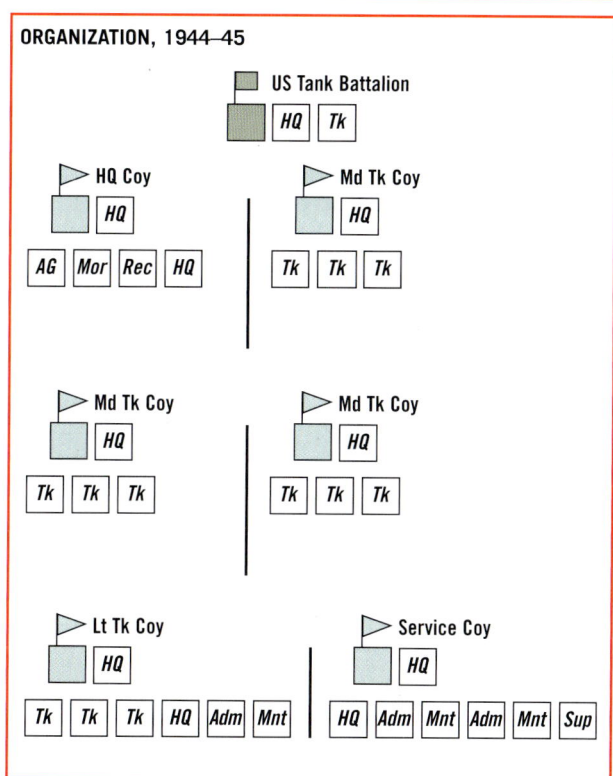

US Tank Battalion, HQ Company	Strength
Company HQ:	
¼-Ton Trucks	2
M3 Halftracks	2
2½-Ton Trucks	1
1-Ton Trailers	1
Battalion Recce Platoon:	
¼-Ton Trucks	5
M3 Halftracks	1
Mortar Platoon:	
M3 Halftracks	1
M21 81mm Mortar Carriers	3
Assault Gun Platoon:	
M3 Halftracks	2
M4 Tanks (105mm/4.1in Howitzer)	3
M10 Ammo Carriers	4

US Tank Battalion, Service Company	Strength
Company HQ:	
¾-Ton Trucks	1
2½-Ton Trucks	3
1-Ton Trailers	3
¼-Ton Trucks	1
Btn Maintenance Platoon:	
¼-Ton Trucks	1
¾-Ton Trucks	1
2½-Ton Trucks	2
1-Ton Trailers	2
M32 ARVs	2
Heavy Wreckers	2
Btn Supply & Transportation Platoon:	
¼-Ton Trucks	1
¾-Ton Trucks	1
2½-Ton Trucks	29
1-Ton Trailers	15
M10 Ammo Trailers	13

US Medium Tank Company	Strength
Company HQ:	
¼-Ton Trucks	2
M4 Tanks	2
M4 Tanks (105mm/4.1in Howitzer)	1
2½-Ton Trucks	1
1-Ton Trailers	1
M3 Halftracks	1
M32 ARVs	1
Tank Platoons x 3	
M4 Tanks	5

▶ **Amphibious operation**
A US Army DUKW amphibious vehicle is used to cross the Danube River alongside a destroyed bridge, May 1945.

INVADING THE REICH

From the Rhine to victory
MARCH–MAY 1945

The Rhine was one of the last major natural obstacles to the Allied advance into the heart of the *Reich*. Remembering the failure of Operation *Market Garden*, Montgomery planned meticulous amphibious and airborne operations to secure bridgeheads across the river, but the first crossing was entirely unplanned, a matter of luck and daring.

ALLIED FORCES WERE now rapidly closing up to the Rhine, and on 7 March Combat Command B of the US 9th Armored Division reached Remagen, seizing the damaged but intact Ludendorff Bridge over the Rhine in one of the first combat operations involving the new M26 Pershing heavy tank. Despite heavy artillery bombardments and repeated *Luftwaffe* air raids against the bridge, three divisions crossed before the weakened structure collapsed on the 17th, by which time pontoon bridges had been completed to allow the offensive to continue.

Away to the north, Montgomery's Twenty-first Army Group was completing preparations for its own crossings of the Rhine. As heavy German resistance was expected, the crossings were to be made in overwhelming force on 23/24 March, with the amphibious assault (Operation *Plunder*) being supplemented by major airborne landings (Operation *Varsity*). Defending German units were generally well understrength with very little armoured support, and many were understandably badly shaken by the preliminary bombardment of 5500 guns along the 35km (22-mile) front. The crossings were made with considerable support from the 'funnies' of 79th Armoured Division and quickly broke through the German defences. The only significant armoured counterattacks were made by 116th Panzer Division against the US Ninth Army's 30th Infantry Division, but these were rapidly broken up by the intervention of 8th Armored Division.

Once across the Rhine, the Allied armies fanned out to secure key objectives. The Ruhr was garrisoned

◀ **M19 Gun Motor Carriage (GMC)**
US Twelfth Army Group / First Army / 639th AA Artillery Battalion*
The M19 was based on a lengthened M24 chassis. The engine was moved behind the driver's compartment and a new open-topped turret with twin 40mm (1.57in) Bofors guns was installed. A total of 342 rounds of 40mm (1.57in) ammunition was carried and there was provision for towing an M28 ammunition trailer with a further 320 rounds. *It is uncertain whether any M19s actually saw action before the end of the war. The type remained in service well into the 1950s and was extensively used in the Korean War in the infantry support role.

Specifications
Crew: 6
Weight: 18 tonnes (17.7 tons)
Length: 5.81m (19ft)
Width: 2.93m (9ft 7in)
Height: 2.96m (9ft 8.5in)
Engine: 163.9kw (220hp) Twin Cadillac 44T4 16-cylinder petrol
Speed: 56km/h (35mph)
Range: 160km (100 miles)
Armament: 2 x 40mm (1.57in) Bofors guns
Radio: SCR508

INVADING THE REICH

by Model's Army Group B (21 divisions totalling 430,000 men) but had been largely isolated from the rest of the *Reich* by intensive Allied bombing, which had wrecked much of the region's road and rail network. The US First and Ninth Armies moved rapidly to encircle Army Group B on 4 April after fighting an especially fierce action against a force of 60 German AFVs crewed by the instructors and students of a *Waffen*-SS Panzer training centre at Paderborn. The sheer size of the newly formed Ruhr Pocket was daunting, and its defenders ranged from elderly *Volkssturm* 'Home Guards' to elite Panzer units (including the Jagdtiger-equipped *Panzerjäger Abteilung* 512).

Whilst many of the *Volkssturm* were only too glad to surrender, the regular units and fanatical Hitler Youth were another matter entirely. Ad hoc anti-tank teams of 12- to 15-year-olds stalked US armour with Panzerfausts whilst others sniped at tank commanders and command posts. At first, determined German units had little difficulty in breaking out as the Allied line ran for over 280km (175 miles) and was initially thinly held by armoured formations whose small infantry units could do little more than hold key points and patrol the perimeter. It was only when the infantry divisions caught up that a proper line could be established. Eventually a total of five infantry divisions, plus a cavalry group, held the line and preparations for the destruction of the pocket could begin.

It took until 12 April to cut Army Group B in two and, although the smaller eastern part of the pocket surrendered the next day, it was not until 21 April that the last units in the western sector finally gave in, bringing the total number of prisoners taken in the region to 325,000.

Mine menace

Montgomery's Twenty-first Army Group moved north-eastwards, with the main thrust directed at Hamburg. The experiences of 7th Armoured Division were typical of those of many other

▼ **Left a little, steady, steady ...**
A Challenger cruiser tank inches across a narrow temporary bridge somewhere in Holland. The Challenger was essentially an enlarged Cromwell mounting a 17pdr in a tall turret. Development was plagued by mechanical problems and it soon became clear that the Sherman Firefly was a much better vehicle. A limited production run of 200 vehicles was authorized, and these were issued on the basis of one per Cromwell troop.

INVADING THE REICH

armoured formations in the last weeks of the war. Although by now the *Panzerwaffe* was reduced to a relative handful of skeleton units, its remaining AFVs were formidable opponents – as a contemporary British report noted: 'The enemy … has brought delaying actions by small bodies of infantry backed by SP guns to a fine art.' In addition to these threats, mines were a constant menace. The conventional anti-tank mines were bad enough, but these were now supplemented by a staggering variety of improvised devices using artillery shells, aircraft bombs and naval mines. Many of these contained far more explosives than normal mines and were capable of inflicting devastating damage.

In one incident, a substantial roadblock was sealed by a large cylindrical concrete block that proved impervious to 75mm (2.9in) HE shellfire from a Sherman Crab. An AVRE was called up and destroyed the concrete obstacle with several rounds from its Petard mortar. As the Crab's flail drum was too wide for it to get through the breach, it was decided that the AVRE would go through and widen

▲ M8 Armoured Car
US Twelfth Army Group / Third Army / 9th Armored Division / 89th Cavalry Reconnaissance Squadron, Mechanized

Over 8500 M8s were built between 1942 and 1943 and the type proved to be a successful escort, patrol and reconnaissance vehicle.

Specifications

Crew: 4
Weight: 8.12 tonnes (8 tons)
Length: 5m (16ft 5in)
Width: 2.54m (8ft 4in)
Height: 2.25m (7ft 5in)
Engine: 82kW (110hp) Hercules JXD 6-cylinder petrol
Speed: 89km/h (55mph)
Range: 563km (350 miles)
Armament: 1 x 37mm (1.5in) M6 gun, plus 1 x 12.7mm (0.5in) AA HMG and 1 x coaxial 7.62mm (0.3in) MG
Radio: SCR508

▲ M3 Halftrack
US Twelfth Army Group / Third Army / 9th Armored Division / 52nd Armored Infantry Battalion

The M3 was deployed in huge numbers – each US armoured division had over 450 of these vehicles.

Specifications

Crew: 2 plus 11 passengers
Weight: 9.3 tonnes (9.15 tons)
Length: 6.34m (20ft 10in)
Width: 2.22m (7ft 3in)
Height: 2.69m (8ft 10in)
Engine: 109.5kW (147hp) White 160AX 6-cylinder petrol
Speed: 72km/h (45mph)
Range: 320km (200 miles)
Armament: 1 x 12.7mm (0.5in) HMG, plus 2 x 7.62mm (0.3in) MGs

the gap from the other side. It had just moved through the breach when it disintegrated in an enormous explosion. Subsequent investigations revealed that the AVRE had been the victim of a naval mine containing over 136kg (300lb) of high explosive, which had also detonated its Petard ammunition and demolition charges. Even when carefully flailing, Sherman Crabs were vulnerable to these devices due to their large blast radius.

On 29 March, the advance was delayed by four well-camouflaged anti-tank guns at Weseke, which had to be taken by an infantry assault using Kangaroos. At Sudlohn, another action had to be fought against a *Panzergrenadier* battlegroup, backed up by 88mm (3.5in) guns. Extensive bomb damage in the town itself held up the advance for some hours whilst 4th Field Squadron RE filled in the craters and strengthened the roads so that they could take the weight of the tanks.

On to Hamburg

The bombed-out ruins of the town of Stadtlohn, a few kilometres north of Sudlohn, had to be cleared of its garrison of two infantry battalions, after which a rickety wooden bridge across a stream was replaced by a Bailey Bridge. The momentum of the advance was maintained by night attacks, however, and by

▲ **All quiet?**
A patrolling M8 armoured car passes a destroyed Stug III assault gun.

1 April, the division had advanced almost 200km (125 miles) in a week.

Before 7th Armoured Division could reach Hamburg, it had to cross the Dortmund–Ems Canal and penetrate the difficult tank terrain of the Teutoburger Wald, a wooded area on an escarpment 40km (25 miles) long and over 1.5km (almost 1 mile) wide. This sector was defended by the cadets of the Hitler Youth Hanover Cadet School and their highly skilled instructors. By 2 April, the town of Ibbenburen at the north-east end of the Teutoburger

Specifications
Crew: 5
Weight: 29 tonnes (28.54 tons) (estimated)
Length: 6.35m (20ft 10in)
Width: 2.9m (9ft 6in)
Height: 2.49m (8ft 2in)
Engine: 447kW (600hp) Rolls-Royce Meteor V12 petrol
Speed: 64km/h (40mph)
Range: 280km (174 miles)
Armament: 1 x 75mm (2.9in) OQF gun, plus 2 x 7.92mm (0.31in) Besa MGs (1 coaxial and 1 ball-mounted in hull front)
Radio: Wireless Set No. 19

▲ **Cromwell Mark VII**
Twenty-first Army Group / Second Army / 7th Armoured Division / 22nd Armoured Brigade / 1 RTR

Later Cromwells such as the Mark VII had their frontal armour increased from 76mm (3in) to 101mm (4in) to improve battlefield survivability.

INVADING THE REICH

▶ AEC Armoured Car Mark III
Twenty-first Army Group / Second Army / XXX Corps / 51st Highland Division / 2nd Derbyshire Yeomanry

The AEC Mark III was the most powerful of all the British-designed armoured cars to see service during the war and equipped the heavy squadrons of armoured car regiments. Main armament was the 75mm (2.9in) OQF gun as fitted to the majority of contemporary Cromwell and Churchill tanks.

Specifications
Crew: 4
Weight: 12.9 tonnes (12.7 tons)
Length: 5.61m (18ft 5in)
Width: 2.69m (8ft 10in)
Height: 2.69m (8ft 10in)
Engine: 116kW (155hp) AEC 6-cylinder diesel
Speed: 66km/h (41mph)
Range: 402km (250 miles)
Armament: 1 x 75mm (2.95in) OQF gun, plus 1 x coaxial 7.92mm (0.31in) Besa MG
Radio: Wireless Set No. 19

◀ Daimler Armoured Car
Twenty-first Army Group / Second Army / XXX Corps / 51st Highland Division / 2nd Derbyshire Yeomanry

A number of Daimlers were modified in 1944–45 by having their turrets removed and pintle-mounted machine guns fitted to produce 'heavy scout cars', which were widely known as 'SODs' – 'Sawn-Off Daimlers'.

Specifications
Crew: 3
Weight: 7.62 tonnes (7.5 tons)
Length: 3.96m (13ft)
Width: 2.44m (8ft)
Height: 2.24m (7ft 4in)
Engine: 71kW (95hp) Daimler 6-cylinder petrol
Speed: 80km/h (50mph)
Range: 330km (205 miles)
Armament: 1 x 40mm (1.57in) 2pdr OQF gun, plus 1 x coaxial 7.92mm (0.31in) Besa MG
Radio: Wireless Set No. 19

▲ Chevrolet C60L
Twenty-first Army Group / Second Army / 7th Armoured Division / 22nd Armoured Brigade / 1st Rifle Brigade

Canadian Military Pattern (CMP) truck production was undertaken on a massive scale with over 900,000 vehicles built between 1940 and 1945. The C60 3-ton 4x4 was produced in the largest numbers, with over 209,000 examples completed.

Specifications
Crew: 1
Weight: 2.1 tonnes (2 tons)
Length: 6.2m (20ft 4in)
Width: 2.29m (7ft 6in)
Height: 3.05m (10ft)
Engine: 71kW (95hp) Ford V8 petrol
Speed: 80km/hr (50mph)
Range: 270km (168 miles)

INVADING THE REICH

▼ US Medium Tank Company

By 1945, many US tank companies had been re-equipped with late-production M4A3E8 Shermans, armed with the 76mm (3in) M1 gun and fitted with the excellent Horizontal Volute Spring Suspension (HVSS). Whilst these tanks were still markedly inferior to the Panther and Tiger, they were a great improvement on earlier Shermans in terms of both firepower and protection.

Company HQ (2 x M4 Shermans, 1 x Sherman with 105mm/4.1in howitzer, 1 x jeep)

Admin, Mess & Supply Section (1 x 2½-ton truck)

Maintenance Section (1 x M3 halftrack, 1 x ARV, 1 x jeep)

1 Platoon (5 x M4 Shermans)

2 Platoon (5 x M4 Shermans)

3 Platoon (5 x M4 Shermans)

Wald was defended by seven companies from the school, who were formidable opponents. They worked in small, widely dispersed groups, showing a particular skill in sniping at tank commanders and platoon sergeants. They inflicted heavy casualties and many of them fought to the death, rather than surrender. After two days of fierce fighting, 7th Armoured Division was ordered to disengage and by-pass the area to continue the advance. It took another two days for the following 52nd and 53rd Infantry Divisions to overcome the cadets.

During the next stage of the advance, *Luftwaffe* fighter-bombers made one of their increasingly rare attacks and an Me 109 was shot down by an AA armoured car of the 11th Hussars. There was another fierce action, this time against the 20th SS Training Division as the advance neared Bremen, followed by a four-day battle against the 2nd Marine Division from Hamburg.

After a brief halt to rest and reorganize, the division moved on. On 16 April, the POW camp *Stalag* XI B, in the woods south-west of Fallingbostel, was liberated and 12,500 prisoners were freed. Soltau was heavily defended and had to be assaulted on the 17th with the assistance of Crocodile and Wasp flamethrower vehicles.

INVADING THE REICH

▲ **Hold it!**
An M26 Pershing is carefully positioned on a raft before being ferried across a river 'somewhere in Germany'.

The advance continued to the Elbe on 20 April, with Wasp flamethrowers also playing a key role in the capture of the town of Daerstorf. By now the RHA Forward Observation Officers had reached the Elbe and were calling down fire on river shipping and rail traffic on the far bank. A night attack by a mixed force of SS, naval ratings, police and Hitler Youth was beaten off on 26 April, followed by further actions against other ad hoc forces that included ships' crews, stevedores, policemen and firemen from Hamburg, submarine crews, *Waffen*- SS, paratroops, *Wehrmacht* soldiers, Hitler Youth and *Volkssturm* 'Home Guards'.

▲ **M24 Chaffee Light Tank**
US Twelfth Army Group / First Army / 3rd Armored Division / 33rd Armored Regiment
The Allied white star was frequently obscured or toned down by tank crews, who believed that it acted as a convenient aiming point for German gunners.

Specifications

Crew: 5
Weight: 18.28 tonnes (18 tons)
Length: 5.49m (18ft)
Width: 2.95m (9ft 8in)
Height: 2.46m (8ft 1in)
Engine: 2 x 82kW (110hp) Cadillac 44T24 V8 8-cylinder petrol
Speed: 55km/h (34mph)
Range: 282km (175 miles)
Armament: 1 x 75mm (2.9in) M6 gun, plus 1 x 12.7mm (0.5in) HMG on AA mount and 2 x 7.62mm (0.3in) MGs (1 coaxial, 1 ball-mounted in hull front)
Radio: SCR508

INVADING THE REICH

These were supported by a powerful assortment of 88mm (3.5in) guns no longer needed for the air defence of Hamburg.

On 28 April, Hamburg itself was brought under artillery bombardment and the following day a deputation from the city came out to discuss surrender. Negotiations went on for some time, but the details were eventually agreed and the surrender was formally signed on 3 May. The division's long journey was not quite finished, though, and units pushed on as far as Kiel, which was occupied by VE Day (8 May 1945).

American advance

Whilst Twenty-first Army Group was completing its advance to the Baltic, US forces were making equally spectacular progress in the south. The First and Ninth Armies moved up to the Elbe, linking up with

▲ 105mm Howitzer Motor Carriage (HMC) M7 (Priest)
US Twelfth Army Group / First Army / 3rd Armored Division / 67th Field Artillery Battalion

All 3490 Priests were armed with 105mm (4.1in) M2A1 howitzers, firing a 14.97kg (33lb) shell to a maximum range of 11,430m (12,505 yards).

Specifications
- Crew: 7
- Weight: 26.01 tonnes (25.6 tons)
- Length: 6.02m (19ft 9in)
- Width: 2.88m (9ft 5in)
- Height: 2.54m (8ft 4in)
- Engine: 298kW (400hp) Continental R975 C1 radial petrol
- Speed: 42km/h (26mph)
- Range: 201km (125 miles)
- Armament: 1 x 105mm (4.1in) M1A2 howitzer, plus 1 x 12.7mm (0.5in) HMG on 'pulpit' AA mount
- Radio: SCR608

▲ M3 Halftrack
US Twelfth Army Group / First Army / 3rd Armored Division / 36th Armored Infantry Regiment

Most M3s carried a 12.7mm (0.5in) Browning heavy machine gun and many added a 7.62mm (0.3in) machine gun on each side of the troop compartment.

Specifications
- Crew: 2 plus 11 passengers
- Weight: 9.3 tonnes (9.15 tons)
- Length: 6.34m (20ft 10in)
- Width: 2.22m (7ft 3in)
- Height: 2.69m (8ft 10in)
- Engine: 109.5kW (147hp) White 160AX 6-cylinder petrol
- Speed: 72km/h (45mph)
- Range: 320km (200 miles)
- Armament: 1 x 12.7mm (0.5in) HMG, plus 2 x 7.62mm (0.3in) MGs

INVADING THE REICH

Soviet forces from Marshal Ivan Konev's 1st Ukrainian Front at Torgau on 25 April. Patton's Third Army fanned out into Czechoslovakia, Bavaria and northern Austria. The US Sixth Army Group skirted Switzerland, moving through Bavaria into Austria and northern Italy. The French First Army advanced through the Black Forest and Baden. The remnants of the German Army Group G fought delaying actions near Nuremberg and Munich but finally surrendered at Haar in Bavaria on 5 May.

The rapid advances of the last days of the war were typified by the experience of elements of the US 71st Infantry Division, which was moving through Bavaria and Austria. Just two days before Germany's final surrender, the 71st was ordered to make contact with elements of the Red Army advancing from Vienna.

On 6 May, the armoured cars of the 71st's cavalry reconnaissance unit were ordered on yet another mission in search of the Soviet forces. Although they failed to find a single Soviet soldier, they did find the headquarters of Army Group *Ostmark* (the former Army Group South), the largest organized field command remaining in the *Wehrmacht*. Its

▶ M29 Weasel
US Twelfth Army Group / Ninth Army / 29th Infantry Division / 121st Engineer Battalion

The amphibious M29 Cargo Carrier 'Weasel' was used in a wide variety of roles including command, radio, ambulance, signal line laying and light cargo vehicle. Its wide tracks gave it exceptional mobility across the deep mud frequently encountered during the winter of 1944/45 which was impassable even by conventional tracked vehicles.

Specifications
Crew: 1 plus 3 passengers
Weight: 2.19 tonnes (2.16 tons)
Length: 4.8m (15ft 9in)
Width: 1.7m (5ft 7in)
Height: 1.8m (5ft 11in)
Engine: 55.9kW (75hp) Studebaker Model 6-170 petrol
Speed: 58km/h (36mph)
Range: 265km (164 miles)

▲ M8 Armoured Car
US Twelfth Army Group / 2nd Armored Division / 82nd Armored Reconnaissance Battalion

The cramped fighting compartment of the M8 meant that much of the crew's kit had to be stowed externally.

Specifications
Crew: 4
Weight: 8.12 tonnes (8 tons)
Length: 5m (16ft 5in)
Width: 2.54m (8ft 4in)
Height: 2.25m (7ft 5in)
Engine: 82kW (110hp) Hercules JXD 6-cylinder petrol
Speed: 89km/h (55mph)
Range: 563km (350 miles)
Armament: 1 x 37mm (1.5in) M6 gun, plus 1 x 12.7mm (0.5in) AA HMG and 1 x coaxial 7.62mm (0.3in) MG
Radio: SCR508

commander, General Lothar Rendulic, was understandably anxious to avoid capture by the rapidly advancing Red Army and surrendered his entire force of four field armies, each numbering around 200,000 men, to the division. As Allied forces celebrated VE Day, very few of them could have foreseen that the end of World War II in Europe marked the beginning of a new kind of war. Most would have been incredulous at the thought that British and US armoured formations would remain in Germany for almost another 50 years as a deterrent force in the Cold War.

Specifications

Crew: 5
Weight: 30.3 tonnes (29.82 tons)
Length: 5.90m (19ft 4in)
Width: 2.62m (8ft 7in)
Height: 2.74m (8ft 11in)
Engine: 372.5kW (500hp) Ford GAA 8-cylinder petrol
Speed: 42km/h (26mph)
Range: 210km (130 miles)
Armament: 1 x 75mm (2.9in) M3 gun, plus 2 x 7.62mm (0.3in) MGs (1 coaxial, 1 ball-mounted in hull front)
Radio: SCR508

▲ M4A3 Sherman Medium Tank
US Twelfth Army Group / 2nd Armored Division / 66th Armored Regiment
The M4's mechanical reliability was off-set by its tendency to burn readily when hit – some crews referred to them as 'Ronsons' after the contemporary cigarette lighter advertised as 'lighting every time'.

Specifications

Crew: 5
Weight: 18.28 tonnes (18 tons)
Length: 5.49m (18ft)
Width: 2.95m (9ft 8in)
Height: 2.46m (8ft 1in)
Engine: 2 x 82kW (110hp) Cadillac 44T24 V8 8-cylinder petrol
Speed: 55km/h (34mph)
Range: 282km (175 miles)
Armament: 1 x 75mm (2.9in) M6 gun, plus 1 x 12.7mm (0.5in) HMG on AA mount and 2 x 7.62mm (0.3in) MGs (1 coaxial, 1 ball-mounted in hull front)

▲ M24 Chaffee Light Tank
US Twelfth Army Group / 2nd Armored Division / 67th Armored Regiment
A total of 4731 Chaffees were produced during 1944/45 and the type was used as the basis for a wide variety of other vehicles including the M19 GMC, the M37 105mm (4.1in) HMC and the M41 155mm (6.1in) HMC.

INVADING THE REICH

▶ Ford WOA
Twenty-first Army Group / HQ Second Army

The Ford WOA was based on a militarized version of a contemporary civilian car chassis and was initially produced as a staff car with saloon bodywork. The type was also produced with a 'utility' body as the WOA2. Total production of both models was approximately 5000 vehicles.

Specifications
Crew: 1
Weight: 2 tonnes (1.96 tons) (estimated)
Length: 4.39m (14ft 5in)
Width: 1.9m (6ft 3in)
Height: 1.78m (5ft 10in)
Engine: 63.32kW (85hp) Ford V8 petrol
Speed: 97km/h (60mph) (estimated)
Range: 280km (175 miles)

▶ Ford WOA2
Twenty-first Army Group / HQ Second Army

The Ford WOA2 entered service in May 1941 and was widely used by all three services as a staff and command car.

Specifications
Crew: 1
Weight: 2.17 tonnes (2.13 tons)
Length: 4.39m (14ft 5in)
Width: 1.9m (6ft 3in)
Height: 1.78m (5ft 10in)
Engine: 63.32kW (85hp) Ford V8 petrol
Speed: 97km/h (60mph) (estimated)
Range: 280km (175 miles)

▲ M26 Pershing Heavy Tank
US Twelfth Army Group / First Army / 3rd Armored Division / 32nd Armored Regiment

Whilst the M26 was appreciated for its protection and firepower, the poor power-to-weight ratio of its 373kW (500hp) Ford engine left it markedly underpowered.

Specifications
Crew: 5
Weight: 41.86 tonnes (41.2 tons)
Length: 8.61m (28ft 3in)
Width: 3.51m (11ft 6in)
Height: 2.77m (9ft 1in)
Engine: 373kW (500hp) Ford GAF V8 petrol
Speed: 48km/h (30mph)
Range: 161km (100 miles)
Armament: 1 x 90mm (3.5in) M3 gun, plus 1 x 12.7mm (0.5in) AA HMG and 2 x 7.62mm (0.3in) MGs (1 coaxial and 1 ball-mounted in hull front)
Radio: SCR508/528

INVADING THE REICH

▲ **Cromwell Mark IV**

Twenty-first Army Group / 7th Armoured Division / 22nd Armoured Brigade / 5th Royal Inniskilling Dragoon Guards

The Cromwell's Christie suspension system and Meteor engine gave it an impressive turn of speed and great manoeuvrability. The engine's governor was frequently decommissioned by the crew to give a higher top speed.

Specifications

Crew: 5
Weight: 27.94 tonnes (27.5 tons)
Length: 6.35m (20ft 10in)
Width: 2.9m (9ft 6in)
Height: 2.49m (8ft 2in)
Engine: 447kW (600hp) Rolls-Royce Meteor V12 petrol
Speed: 64km/h (40mph)
Range: 280km (174 miles)
Armament: 1 x 75mm (2.9in) OQF gun, plus 2 x 7.92mm (0.31in) Besa MGs (1 coaxial and 1 ball-mounted in hull front)
Radio: Wireless Set No. 19

▲ **M4A3E8 Sherman Medium Tank**

US Twelfth Army Group / 2nd Armored Division / 67th Armored Regiment

Although the 76mm (3in) gun of late-model US Shermans had a markedly better anti-tank performance than the earlier 75mm (2.9in), it was outclassed by the 17pdr mounted by the Firefly.

Specifications

Crew: 5
Weight: 33.7 tonnes (33.16 tons)
Length: 7.57m (24ft 10in)
Width: 3m (9ft 10in)
Height: 2.97m (9ft 9in)
Engine: 372.5kw (500hp) Ford GAA 8-cylinder petrol
Speed: 42km/h (26mph)
Range: 161km (100 miles)
Armament: 1 x 76mm (3in) M1A1 gun, plus 2 x 7.62mm MGs (1 coaxial and 1 ball-mounted in hull front)
Radio: SCR508

APPENDICES

The development of US armour

By 1918, the newly formed American Tank Corps was developing as a formidable force. Massive production of a wide range of AFVs was planned, including 4400 6-ton M1917 tanks (the US version of the Renault FT-17) and 1450 Mark VIII heavy tanks. Ford were to supplement these with 15,000 machine-gun carriers designated 'two-man tanks' and a further 1000 'three-man tanks' armed with a 37mm (1.5in) gun.

THE TANK CORPS was to have a number of brigades based on the contemporary British practice, each with two light battalions and a single heavy battalion, but by the end of the war, only three light battalions were operational. This ambitious programme was drastically cut back following the 1918 Armistice and only about 1000 M1917s, 100 Mark VIIIs and 15 Ford two-man tanks plus a single example of the Ford three-man tank were actually built. The National Defense Act of 1920 abolished the Tank Corps, bringing the remaining armoured units (four tank battalions) under the control of the Chief of Infantry, with their mission defined as being 'to facilitate the uninterrupted advance of the rifleman in the attack'. The first signs of change came after the US Secretary of War visited the British Experimental Mechanised Force manoeuvres in 1927 and ordered the formation of a similar force. In the following year this assembled at Fort Meade, Maryland, equipped with Mk VIIIs and M1917s.

So much unreliable, obsolete equipment caused dire problems and the force disbanded after three months, but re-formed at Fort Eustis, Virginia, in 1930 with its heavy tank battalion replaced by a second infantry battalion. This time it demonstrated the advantages of mechanization to the rest of the US Army and won the support of its new chief of staff, General Douglas MacArthur. Although the Mechanized Force was officially disbanded in 1931, most of its units were transferred to Fort Knox, Kentucky, to form the nucleus of the 1st Cavalry Regiment (Mechanized) in 1932, but the unit remained incomplete until 1939 when it was finally brought up to its full strength of:
- Two mechanized cavalry regiments (totalling 112 light tanks).
- A motorized artillery regiment.

The Panzer victories in Poland and France prompted a massive expansion of US armour – in July 1940 the Armored Force was formed under Brigadier-General Adna Chaffee with an initial establishment of two armoured divisions and an independent tank battalion. Each division included:
- An armoured brigade of two light tank regiments (each of three battalions of M3 light tanks) plus a two-battalion medium tank regiment with M3 medium tanks.
- A two-battalion artillery regiment with self-propelled 105mm (4.1in) howitzers.
- A towed artillery battalion.
- A two-battalion motorized infantry regiment.

▶ **M2A4 Light Tank**
Desert Training Center, California-Arizona Maneuver Area (DTC-CAMA)
In 1941–42 the then Major-General George Patton commanded the newly established Desert Training Center, at which US armoured warfare doctrine was developed. This M2A4 served as Patton's personal tank at the DTC.

Specifications

Crew: 4
Weight: 11.6 tonnes (11.4 tons)
Length: 4.43m (14ft 6in)
Width: 2.47m (8ft 1in)
Height: 2.64m (8ft 8in)
Engine: 186.25kW (250hp) Continental W-670-9A 7-cylinder petrol
Speed: 56km/h (36mph)
Range: 110km (70 miles)
Armament: 1 x 37mm (1.5in) M20 gun, plus 5 x 7.62mm (0.3in) MGs
Radio: n/k

APPENDICES

- A reconnaissance battalion.
- An engineer battalion.

Although each division officially had 381 tanks, there were only 44 serviceable tanks in the entire US Army and vast industrial resources were committed to AFV production to make up the required numbers. The total annual figures for the early 1940s were:
- 1940 – 330
- 1941 – 4052
- 1942 – 24,997
- 1943 – 29,487

Three more armoured divisions were raised before America entered the war in December 1941 and a further 11 followed by 1943. These divisions were extensively reorganized between 1941 and 1943 – the numbers of tanks were reduced whilst the infantry component was increased. The old brigade structure was abolished and replaced by two (and finally three) Combat Command HQs which came directly under the Divisional HQ and could be used to control any combination of armour, artillery and infantry as required. Thus, by 1943, the armoured divisions' components were:

- Combat Commands A, B and C (CCA, CCB and CCC).
- A reconnaissance battalion.
- Three tank battalions with a total of 219 tanks. (At first, many were equipped with M3 Mediums, but these were being replaced by M4 Mediums as they became available).
- Three armoured infantry battalions carried in M2 or M3 halftracks.
- Three artillery battalions equipped with self-propelled 105mm (4.1in) howitzers.
- An engineer battalion.

In addition to the armoured divisions, almost 50 additional tank battalions were deployed independently in the infantry support role by 1943.

Tank destroyer battalions

The Armored Force was supported by a separate Tank Destroyer (TD) Force which was raised in 1941 by Lieutenant-Colonel Andrew D. Bruce. The basic TD unit was the 36-gun battalion, although TD groups each of two battalions and TD brigades with two groups apiece were sometimes formed. At first, their equipment was basic, including the unarmoured 37mm Gun Motor Carriage (GMC) M6, a 37mm (1.5in) gun on a 15cwt truck, and the 75mm GMC M3, which was the old French 75mm (2.9in) field gun on the M3 halftrack. These were used in action in Tunisia, although by the end of the campaign, the much more formidable 76mm (3in) GMC M10 was entering service. Although the vast US industrial base could supply almost unlimited quantities of AFVs, in the end everything depended on just how quickly their inexperienced crews and commanders could learn to cope with the realities of armoured warfare.

▲ M3 Halftrack
Desert Training Center, California-Arizona Maneuver Area (DTC-CAMA)
This heavily modified M3 was fitted with an armoured roof and a radio for its role as Patton's armoured command vehicle at the Desert Training Center. Patton gained a reputation as a ruthlessly efficient trainer, insisting that 'A pint of sweat will save a gallon of blood'.

Specifications
Crew: 2, plus up to 11 passengers
Weight: 9.1 tonnes (8.95 tons)
Length: 6.16m (20ft 3in)
Width: 1.96m (6ft 5in)
Height: 2.3m (7ft 6in)
Engine: 109.51kW (147hp) White 160AX
6-cylinder petrol
Speed: 72km/h (45mph)
Range: 320km (200 miles)
Armament: 1 x 7.62mm (0.3in) MG
*NB Data is given for standard early-production M3 halftrack

APPENDICES

Combat trials – M26 Pershing heavy tank

In early 1945, the US Army finally recognised that there was a desperate need for a heavier tank than the Sherman, which still equipped the vast majority of US (and many British) armoured units.

The M26 Pershing heavy tank was the result of a programme to produce a successor to the M4 Sherman. In May 1942, the US Ordnance Department received orders to begin development of a new medium tank that would eliminate some of the shortcomings of the M4.

Although the experimental T26 (essentially the same design as the M26) could probably have entered production in late 1943 or early 1944, wrangling over the necessity for such a heavy tank delayed acceptance of the design until November 1944. There was still strong opposition to the whole heavy tank concept until the Ardennes Offensive, when German Tigers and Panthers inflicted severe losses on Sherman-equipped units.

This was the turning point and a batch of 20 M26 Pershings, including one of the two experimental M26A1E2 'Super Pershings' to be completed by the end of the war, were sent to Europe for combat trials. The M26A1E2 was armed with the 90 mm (3.5in) L/70 T15E1 high-velocity gun firing a 7.58kg (16.7lb) T44 HVAP round at 1,170 m/s (3,850 ft/s) and was capable of penetrating 220mm (8.5in) of rolled homogenous armour (RHA) at 30 degrees at 914m (1000 yards).

Pershings in action

The Pershings were issued in two batches of 10 vehicles each, one being assigned to 3rd Armored Division and the second to 9th Armored Division. Of these, a total of three were lost in action, two of them by 3rd Armored Division.

The best documented of these occurred on 6th March 1945 in an action at Niehl, north of Cologne, when a single shot from a Panzerjaeger Nashorn of *Panzer Jaeger Abteilung 93* penetrated the frontal armour of one of 3rd Amored Division's Pershings at a range of only 250m (273.4 yards). The American crew were incredibly lucky as the German round struck the lower left frontal armour, passed between the driver's legs and under the turret basket where it started a fire. All the crew were able to bale out safely before the fire caused a major ammunition explosion.

The sole M26A1E2 was issued to 3rd Armored Division's 33rd Armored Regiment. It received extensive field modifications to improve its armour (master-minded by Lt. Belton Cooper) before it went into action on 4th April between the River Weser and Northheim. Lt. Cooper wrote that:

'Some of the German units that had fallen back from the bridgehead set up a few isolated strong points along our route. One such position on a wooded hill ... opened fire as the column passed. The Super M26, in the forward part of the column, immediately swung its turret to the right and fired an armor-piercing shot toward an object on the forward slope of a wooded hill about fifteen hundred yards away [over three-quarters of a mile]. A blinding flash of sparks accompanied a tremendous explosion as debris shot fifty feet into the air ... The unknown object was a tank or self-propelled gun; had it been a half-track or other vehicle, the flash would not have been as large ... The rest of the column let go with a deluge of tank and automatic weapons fire, and the Germans soon broke off the action ... we didn't know what the Super M26 hit ... no one was anxious to go over and check it out.'

Super Pershing versus Tiger II

The Super Pershing was next in action at Dessau on 21st April, commanded by SSgt Joe Maduri, a veteran of 10 months' continuous combat. The attack on Dessau was initially delayed by numerous concrete anti-tank obstacles which had to be laboriously cleared before the tanks could reach the city centre. Maduri's Super Pershing, accompanied by supporting infantry, reached a road junction and began to turn

APPENDICES

right when it came under fire at a range of roughly 550m (600 yards) from a Tiger II lying in ambush.

Fortunately for Maduri and his crew, the German gunner's aim was terrible and the 88mm (3.45in) shot missed. Maduri's gunner, Cpl John 'Jack' Irwin, only 18 years old, reacted almost instantly with an HE round that struck the Tiger's frontal armour and ricocheted skywards before exploding harmlessly. The Super Pershing had been loaded with an HE round only because Irwin had been expecting 'soft' urban targets, such as barricades, personnel and anti-tank guns. 'AP!' he shouted to his loader, to indicate that an armour-piercing round would be next.

Maduri and crew then felt a thud on the turret. It was never established if this shot came from the Tiger, or from some other anti-tank weapon, but no serious damage was done by the glancing blow. Irwin quickly got in a second shot as the Tiger advanced over a pile of rubble, briefly exposing its thinly armoured underside. The 90mm (3.5in) AP round penetrated the Tiger's belly plates, detonating the ammunition in tremendous explosion which blew off the turret. The whole engagement took less than 30 seconds.

Maduri's crew spent the rest of the day fighting off the German infantry anti-tank teams who stalked their tank with panzerfausts. The next day, they were back in action, clearing the final German held areas of the city when they encountered a Panzer Mk V 'Panther', which they immobilised with a first round hit on a drive sprocket. Their second round penetrated the side armour, causing a massive ammunition explosion which wrecked the vehicle.

▲ **M26 Pershing prototype**
Although the Pershing's armament and armour were a welcome improvement to those of the Sherman, the early models were distinctly under-powered and prone to breakdowns.

APPENDICES

US armoured divisions

All US armoured divisions used the same basic badge of a triangle divided into three with the top yellow, bottom left blue and bottom right red. The centre bore a stylized armoured vehicle track in black under a red lightning bolt. The sole distinction between badges was the divisional number in black on the yellow section.

1st Armored Division (Nickname: *Old Ironsides*)

The division can trace its origins back to US experiments in mechanized warfare of the early 1930s utilizing the 7th Cavalry Brigade. On 15 July 1940, this formation was expanded to form the 1st Armored Division. As the US Army had no previous experience with armoured formations of this size, it took considerable time to develop a workable structure. (In the process, the division was subjected to a total of six separate reorganizations between 1940 and 1945!) After extensive training, the division deployed to the UK in May 1942 and prepared for Operation *Torch*, the invasion of north-west Africa. It landed near Oran in November 1942 and fought throughout the Tunisian campaign, gaining hard-won experience of the realities of armoured warfare at Kasserine Pass in February 1943. After reorganizing in Morocco, the division landed in Italy in October 1943, taking part in the assaults on the Winter Line, the Anzio landings and the liberation of Rome. It remained in Italy for the rest of the war, where it often had to fight on a 'single-tank front' over some of the world's most difficult terrain for armoured operations. In April 1945, the division took part in the final Allied offensive into the Po Valley, where it could once again operate as a complete formation. It took Milan on 30 April and had advanced 370 km (230 miles) in 19 days by the time that German forces in Italy surrendered on 2 May.

2nd Armored Division (Nickname: *Hell on Wheels*)

The 2nd Armored Division was formed at Fort Benning, Georgia, on 15 July 1940. Although elements of the division took part in Operation *Torch*, it did not see action as a complete formation until the invasion of Sicily. It was then transferred to the UK to prepare for Operation *Overlord* and landed in Normandy on 9 June 1944, initially operating in the Cotentin Peninsula before taking part in Operation *Cobra*. In the autumn of 1944, the division broke through the Siegfried Line and advanced to the River Roer, but had to hastily redeploy to Belgium to help contain the Ardennes Offensive and regain the territory lost during the German attack. After crossing the Rhine in March 1945, 2nd Armored Division was the first US formation to reach the Elbe on 11 April, where it was ordered to halt. In July, it became the first US division to enter Berlin.

3rd Armored Division (Nicknames: *Spearhead Division* or the *Third Herd*)

The division was established on 15 April 1941 and deployed to the UK in September 1943 in preparation for Operation *Overlord*. After pre-invasion training, it landed in Normandy on 24 June 1944, taking part in the battle for St Lô and subsequently helping to trap the German Seventh Army in the Falaise Pocket.

By early September, the division had advanced into Belgium, cutting off 40,000 German troops at Mons. On 10 September, it claimed to fire the first US field artillery shell onto German soil and subsequently assaulted the Siegfried Line before participating in the Battle of Hürtgen Forest.

The 3rd Armored Division was heavily involved in the Allied counterattacks following the Ardennes Offensive, after which it resumed its advance into the Rhineland, taking Cologne and Paderborn before ending the war at Dessau.

4th Armored Division

The division was formed on 15 April 1941 based on a cadre drawn from the 1st Armored Division and deployed to the UK early in 1944. It landed in Normandy on 11 July and took part in Operation *Cobra*, advancing into Lorraine by September. In December, during the Ardennes Offensive, the division covered 240 km (150 miles) in 19 hours to raise the siege of Bastogne, subsequently closing up to the Rhine. After crossing the Rhine on 24/25 March 1945, the division rapidly advanced through south-eastern Germany, ending the war in Czechoslovakia.

5th Armored Division (Nickname: *Victory*)

The 5th Armored Division was established on 10 October 1941 and deployed to the UK in February 1944. It landed in Normandy on 24 July, taking Le Mans on 8 August before advancing to Argentan to help form the Falaise Pocket.

It moved on to liberate Luxembourg City on 10 September, and on the following day, one of its reconnaissance patrols were the first Allied troops to cross the German frontier. After heavy fighting in the Hürtgen Forest, the division was temporarily placed in Twelfth Army Group reserve. In March 1945, the division closed up to the Rhine, crossing the river at Wesel on 30 March. By 12 April, it had reached the Elbe at Tangermünde and was only 72 km (45 miles) from Berlin. Unable to advance following Eisenhower's order to halt at the Elbe, the division took part in mopping-up operations in the Ninth Army's sector until VE Day.

APPENDICES

6th Armored Division (Nickname: *Super Sixth*)

The 6th Armored Division was formed on 15 February 1942 based on a cadre from 2nd Armored Division. It deployed to the UK in February 1944, landing in Normandy in July 1944 where it took part in Operation *Cobra*, before being sent to hold the siege lines around the German garrison in Lorient. In November, the division moved to the Saarland, reaching the German border on 6 December and setting up defensive positions around Saarbrücken.

The division was heavily engaged in the counterattacks around Bastogne following the Ardennes Offensive before closing up to the Rhine in March 1945. It crossed the river on 25 March and advanced to take Frankfurt and Mühlhausen.

On 11 April, the division liberated Buchenwald concentration camp and moved on to take Leipzig. By 15 April, it had advanced as far as the River Mulde, where it was ordered to halt and await the arrival of the Red Army.

7th Armored Division (Nickname: *Lucky Seventh*)

The division was formed on 1 March 1942 and deployed to the UK in June 1944. It landed in Normandy in August, coming under the command of the US Third Army, and liberated Chartres, before advancing across the Seine to take Château-Thierry and Verdun. It then became bogged down in fierce fighting around Metz before being transferred to the US Ninth Army on 25 September for operations to protect the right flank of the salient formed by Operation *Market Garden*.

In response to the Ardennes Offensive, the division was transferred to the US First Army and rushed to St Vith where it fought fierce delaying actions for almost a week before being ordered to evacuate the town on 23 December. It was then heavily engaged in the Allied counterattacks against the 'Bulge' created by the offensive, finally retaking St Vith on 23 January 1945.

In March 1945, 7th Armored Division crossed the Rhine and took part in the breakout from the Remagen bridgehead, helping to surround the Ruhr Pocket. It then played a key role in the destruction of the pocket, forcing the surrender of LIII Panzer Corps on 16 April, before being transferred to Twenty-first Army Group for operations on the Baltic coast in the final days of the war.

8th Armored Division (Nicknames: *Iron Snake, Thundering Herd, Tornado*)

The division was formed on 1 April 1942, deployed to the UK in November 1944 and moved to France in January 1945. Its first major actions were fought around the River Roer as part of Operation *Grenade*. It then advanced to the Rhine, fighting a fierce action against 130th Panzer Division near Rheinberg.

On 24 March, the division crossed the Rhine as part of Operation *Plunder* and advanced as far as Paderborn and Sennelager before being diverted to assist in the reduction of the Ruhr Pocket on 3 April. After 10 days of intensive combat, 8th Armored was ordered eastwards to clear the area around the Harz Mountains where it operated until the end of the war.

9th Armored Division (Nickname: *Phantom Division*)

The division was formed on 15 July 1942 and deployed to the UK in September 1944 before landing in Normandy towards the end of the month. It was initially sent to a quiet sector of the front on the Luxembourg-German border, but was caught by the full weight of the Ardennes Offensive, with its units fighting in scattered groups at St Vith and Bastogne.

After a brief rest period, the division advanced to the Rhine, seizing the Ludendorff Bridge at Remagen on 7 March 1945 and establishing a bridgehead across the river before moving on to take Frankfurt and surround the Ruhr Pocket. During the following month, it advanced into eastern Germany, by-passing Leipzig and moving up to the River Mulde. As the war ended, the division was en route to reinforce US troops in Czechoslovakia.

10th Armored Division (Nickname: *Tiger Division*)

The 10th Armored Division was formed on 15 July 1942, deploying to France in September 1944. It fought its first actions in the Metz area in November, before being caught up in the Ardennes Offensive, defending Bastogne, Noville and Bras. In February 1945, it cleared the Saar-Moselle Triangle before crossing the Rhine at Mannheim on 28 March and advancing into Bavaria. The 10th took Oberammergau and reached Innsbruck by VE Day.

11th Armored Division (Nickname: *Thunderbolt*)

The division was established on 15 August 1942, deploying to France in December 1944. It was immediately involved in heavy fighting to contain the Ardennes Offensive and in the subsequent actions to regain the 'Bulge'. In March 1945, the division helped to clear the Saar-Moselle-Rhine Pocket, before crossing the Rhine and advancing into Bavaria. By 14 April, it had taken Coburg and Bayreuth and was moving on Linz, which was captured on 5 May. Elements of the division made contact with Soviet forces on 8 May, the day before the official end of the war in Europe.

12th Armored Division (Nickname: *Hellcat Division*)

The division was formed on 15 September 1942 and deployed to France on 11 November 1944. After taking heavy losses in its first attacks against the German Rhine bridgehead at Herrlisheim, the 12th took part in the reduction of the Colmar Pocket. It was then engaged in operations to clear the Saar Palatinate before crossing the Rhine at Worms on 28 March. In the final month of the war the division swept across Bavaria and into Austria.

177

APPENDICES

13th Armored Division
(Nickname: *The Black Cats*)

The division was formed on 15 October 1942, deploying to France at the end of January 1945. In April 1945, it took part in the reduction of the Ruhr Pocket before advancing into Bavaria and crossing the Danube. On 2 May, the division moved into Austria, taking Branau am Inn, and was preparing to advance further into Austria when the war in Europe ended.

14th Armored Division
(Nickname: *The Liberators*)

The division was formed on 15 November 1942 and deployed to France in October 1944. It advanced into Alsace and was on the point of breaking through the Siegfried Line when it was withdrawn in response to the Ardennes Offensive. The 14th was caught up in the subsidiary German offensive Operation *Nordwind* in January 1945 and suffered heavy tank losses in almost two weeks of fierce fighting.

In April, the division crossed the Rhine and moved into Bavaria, earning its nickname by accomplishing the liberation of approximately 200,000 Allied prisoners of war (POWs) from several camps including *Stalag* VII-A, Germany's largest POW camp. The 14th advanced as far as Mülhdorf am Inn and established bridgeheads across the River Inn, where it was ordered to halt, ending its combat operations on 2 May.

16th Armored Division
(Nickname: *Armadillo*)

The division was formed on 15 July 1943, deploying to France in February 1945. Its sole major combat actions were fought between 6 and 8 May 1945 when it advanced into Czechoslovakia to take the Skoda weapons factory complex in Pilsen.

20th Armored Division
(Nickname: *Armoraiders*)

The division was formed on 15 March 1943 and deployed to France in February 1945. On arrival, it was assessed as requiring substantial additional training before being committed to action. As a result, only its 27th Tank Battalion was involved in any significant combat (29–30 April) when it was detached to support the 42nd Infantry Division's attack on Munich.

British armoured divisions

British experience of armoured warfare dated back to 1916 and this, coupled with the tactics developed by the Experimental Mechanised Force in the inter-war years, gave the British Army a head start in developing effective armoured formations. Much of this advantage was lost as financial crises of the period drove successive governments to cut back defence spending until the risks of war became too obvious to ignore. Even then resources were poorly allocated and the lessons of the pre-war period had to be painfully relearned.

1st Armoured Division

The 1st Armoured Division was rushed to France in 1940 after the scale of the German offensive became apparent. The formation was 'robbed' of some of its units in a vain attempt to defend Calais and Boulogne and the remnants fought south of the Somme until evacuated on 16 June.

The division remained on anti-invasion duties in the UK until August 1941, when it was sent to North Africa, arriving in November. It fought in most of the major actions of the desert war, including both battles of El Alamein, before being transferred to Italy in May 1944. It took part in the assault on the Gothic Line before ceasing to be an operational unit in October 1944. It was finally disbanded on 1 January 1945.

2nd Armoured Division

The 2nd Armoured Division was formed in December 1939, but remained in the UK until early 1941, when it deployed to North Africa. Initially guarding the lines of communications in Cyrenaica, it lost its 1st Armoured Brigade which was sent to Greece in a vain attempt to halt the German invasion. Most of the division's remaining units were captured in April 1941, although some managed to escape and were evacuated from Tobruk. The division was disbanded on 10 May 1941 and was not re-formed.

APPENDICES

6th Armoured Division

The 6th Armoured Division formed in the UK on 12 September 1940, but did not deploy abroad until November 1942 when it took part in Operation *Torch*. Operating under command of the British First Army in Tunisia, it spearheaded the final drive on Tunis.

Following the end of the North African campaign, the division was sent to Italy where it came under the command of the Eighth Army (and later the US Fifth Army), taking part in the fighting around Monte Cassino, the assaults on the Gothic Line and the final offensive in April 1945.

7th Armoured Division

The 7th Armoured Division began life as the Mobile Force at Mersa Matruh in Egypt in 1938. It was soon retitled the Mobile Division (Egypt) and came under the command of Major-General Percy Hobart whose relentless training turned it into a formidable armoured force. In February 1940 it became the 7th Armoured Division and adopted the 'Desert Rat' insignia. A few months later it was in action against Italian forces, going on to fight in every major action in the desert war.

The division fought in the early stages of the Italian campaign, before returning to the UK to prepare for Operation *Overlord*. It landed in Normandy on the afternoon of D-Day, taking part in Operations *Goodwood*, *Spring* and *Bluecoat* before the breakout from the beachhead. It then moved up to the Seine before spearheading the advance into Belgium. In 1945, the division crossed the Rhine in Operation *Plunder*, capturing Hamburg and Kiel at the end of the war.

8th Armoured Division

The 8th Armoured Division was formed in November 1940 and eventually deployed to Egypt. Troop shortages meant that it never received its lorried infantry brigade, and although consideration was given to using its remaining elements in a deep penetration role in the pursuit after El Alamein, other units were used. The division was disbanded in Egypt on 1 January 1943.

9th Armoured Division

The 9th Armoured Division was raised on 1 December 1940 and served as a training and trials formation in the UK until disbanded on 31 July 1944.

10th Armoured Division

The 10th Armoured Division was raised in August 1941 from the 1st Cavalry Division which had been based in Palestine. It fought at Alam el Halfa and El Alamein, remaining in North Africa after the end of the desert war. It was disbanded in Egypt on 15 June 1944.

11th Armoured Division

The division formed in Yorkshire in March 1941 under the command of Major-General Percy Hobart who initiated an intensive training programme based on his experience with the 7th Armoured Division. The formation was engaged in training and home defence duties until July 1944 when it deployed to Normandy. It took part in Operations *Epsom* and *Goodwood*, before capturing Antwerp on 4 September. In March 1945, it crossed the Rhine, advancing deep into Germany and liberating Bergen-Belsen concentration camp on 15 April. The 11th took Lübeck on 2 May and spent the immediate post-war period administering the province of Schleswig Holstein until the division's disbandment in January 1946.

42nd Armoured Division

On 1 November 1941, the 42nd (East Lancashire) Infantry Division was formally converted to an armoured division, becoming the 42nd Armoured Division. It remained in the UK as a home defence/training formation until its disbandment in October 1943.

79th Armoured Division

In April 1943, 79th Armoured Division was formed from both Royal Engineers and Royal Armoured Corps units to provide specialised AFVs for dealing with the German fortifications of the 'Atlantic Wall' along the northern French coast. These specialized AFVs proved highly effective during the Normandy landings and the subsequent advance across Europe.

By 1945, the formation was the largest armoured division in the British Army – with 1,050 tracked vehicles, it was three times the size of a normal armoured division. The 79th Armoured Division was disbanded on 20th August 1945, but it had established the need for an armoured engineering capability and 32 Engineer Regiment continues to fulfil this role in the British Army, bearing the Bull's Head badge inherited from the division.

Guards Armoured Division

After its formation on 17 June 1941, the Guards Armoured Division was stationed in the UK on home defence and training duties. It deployed to Normandy on 26 June 1944, taking part in Operations *Goodwood* and *Bluecoat*.

After the breakout from Normandy, it liberated Brussels on 3 September. The division then spearheaded XXX Corps' attack during Operation *Market Garden*, linking up with US airborne forces at Eindhoven and Nijmegen, but failing to cut through strengthening German defences in time to relieve the 1st Airborne Division at Arnhem. During the Ardennes Offensive, the division was deployed to defend the line of the Meuse against a possible German breakthrough. In February/March 1945, it took part in Operation *Veritable* and the subsequent advance through northern Germany. In June 1945, the formation was re-formed as an infantry division, becoming the Guards Division.

APPENDICES

Other Allied armoured divisions

The three Free French armoured divisions raised in 1943 were all organized on US lines, although their constituent units retained traditional titles (such as *12e Régiment de Chasseurs d'Afrique*). The Polish, Canadian and South African divisions all followed British organization.

French 1st Armoured Division (*1re Division Blindée* – 1re DB)

The division formed at Mascara, Algeria, on 1 May 1943 under the command of Brigadier Jean Touzet de Vigier. In August 1944, it formed part of General de Lattre de Tassigny's French First Army for Operation *Dragoon*, the invasion of southern France. 1re DB took part in the liberation of Toulon and Marseilles, before advancing up the Rhone Valley to the Rhine. In 1945 it spearheaded French First Army's advance through southern Germany and was the first French formation to reach the River Danube.

French 2nd Armoured Division (*2e Division Blindée* – 2e DB)

Initially known as the 2nd Light Division, 2e DB was formed in August 1943. It deployed to Normandy on 1 August 1944 under the command of Major-General Philippe Leclerc. It took part in Operation *Cobra* as part of Patton's Third Army, during which it largely destroyed 9th Panzer Division, before liberating Paris on 25 August.

2e DB then advanced into Lorraine and Alsace, destroying 112th Panzer Brigade and liberating Strasbourg on 23 November. In February 1945, the division was sent to assist in the reduction of the Royan Pocket on the French Channel coast, which surrendered on 18 April. The division was then redeployed to southern Germany where it operated against the remnants of Army Group G, ending the war in Berchtesgaden.

French 5th Armoured Division (*5e Division Blindée* – 5e DB)

The original 2nd Armoured Division was formed on 1 May 1943, but was redesignated as the 5th Armoured Division on 16 July 1943 to allow 2nd Free French Division to become 2nd Armoured Division. Originally comprising a tank brigade and a support brigade, 5e DB was re-equipped and reorganized along US lines with three combat commands which were commonly detached to support French infantry divisions.

The division deployed to France in September 1944 and took part in the battles for Belfort and the reduction of the Colmar Pocket. It then went into reserve before supporting the French crossing of the Rhine in March and participating in the final campaign in Germany.

Polish 1st Armoured Division

The Polish 1st Armoured Division (*1 Dywizja Pancerna*) was formed in Scotland in February 1942 under the command of General Stanislaw Maczek. It deployed to Normandy at the end of July 1944 and played a key role in the Battle of Falaise, in which it sealed off the Falaise Pocket, trapping 80,000 German troops. Following the Allied breakout from Normandy, the division advanced along the Channel coast, liberating Ypres, Ghent, Passchendaele and Breda. In early 1945, it fought its way along the Dutch-German border, before advancing on the German naval base of Wilhemshaven, which was taken on 6 May.

4th Canadian Armoured Division

The division was formed in Canada early in 1942 and deployed to the UK later that year. It landed in Normandy at the end of July 1944, taking part in the Battle of Falaise. After the Allied breakout from Normandy, the division advanced rapidly, crossing the Seine and liberating St Omer on 5 September before crossing into Belgium. 4th Armoured Division was then committed to the long struggle to clear the Scheldt estuary to allow Allied supply convoys to use Antwerp docks, which dragged on until 8 November. After refitting during the winter of 1944/45, the division took part in Operation *Veritable*, taking Xanten after a fierce fight against 116th Panzer Division. Following the Rhine crossings, it operated in the eastern Netherlands before advancing into Germany, taking Oldenburg on 5 May.

5th Canadian Armoured Division (Nickname: *The Mighty Maroon Machine*)

In 1941, the 1st Canadian Armoured Division was redesignated as the 5th Canadian Armoured Division and deployed to the UK in November of that year. In November 1943, it replaced the British 7th Armoured Division in Italy and took part in the breaching of the Hitler and Gothic Lines. In January 1945, the division was transferred to Belgium, joining First Canadian Army for operations in the eastern Netherlands and north-west Germany.

6th South African Armoured Division

This was the first ever South African armoured division, formed on 1 February 1943 from elements of the 1st and 2nd South African Infantry Divisions. After training in Egypt, the division deployed to Italy in April 1944, taking part in the capture of Florence on 4 August and the breaching of the Gothic Line, as well as the spring offensive of 1945.

French and British AFV production

The new generation of French AFVs included formidable designs such as the Somua and the Char B1 bis, but many of the factories building them were reliant on antiquated machinery and archaic working practices. (In the Hotchkiss factory, most components were hand-finished with files, as they had been in the nineteenth century.) Matters improved after the outbreak of war brought a greater sense of urgency, but many promising designs, such as the Renault G1 R heavy tank, were still at the prototype stage at the time of the French surrender.

FRENCH TANKS, MODEL	pre Sep '39	1939	1940	Total
Char 2C	10	–	–	10
Renault FT-17	1580	–	–	1580
NC27	36	–	–	36
Char D1	160	–	–	160
Char D2	50	–	50	100
R-35	1070	200	331	1601
FCM-36	100	–	–	100
Char B1	163	42	200	405
AMR-33	123	–	–	123
AMR-35	167	–	–	167
ZB	16	–	–	16
ZT2	–	–	10	10
ZT3	10	–	–	10
ZT4	–	–	40	40
AMC-34	12	–	–	12
AMC-35	22	23	5	50
Hotchkiss H-35	640	130	322	1092
Somua S-35	270	50	110	430
AMD White TBC	86	–	–	86
AMD Laffly 50	98	–	–	98
AMD Laffly 80	28	–	–	28
AMD Laffly S15 TOE	45	–	–	45
AMD Berliet VUDB	32	–	–	32
AMD Panhard 165/175	30	–	–	30
AMD Panhard 178	219	69	239	527
AMC Schneider P16	96	–	–	96
Total	5063	514	1307	6884

In 1936, the total number of tanks held by the British Army was 375 (209 lights and 166 mediums) of which 304 were officially categorized as obsolete. The only 'modern' vehicles were the 69 Light Tanks Marks V and VI, plus two experimental medium tanks. Although a massive effort went into British tank production from the late 1930s, the types selected for production were often hopelessly inadequate as combat vehicles. It was only at the very end of the war that the first prototypes of a tank able to match German types, the Centurion, were completed.

BRITISH TANKS, MODEL	Total
Mk I, Matilda I (A11)	140
Mk II, Matilda II (A12)	2987
Mk III, Valentine	8275
Mk IV, Churchill (A22)	–
Churchill Mk I	303
Churchill Mk II	1127
Churchill Mk III	675
Churchill Mk IV	1622
Churchill Mk V	241
Churchill Mk VI	200
Churchill Mk VII	1600
Churchill Mk VIII	1600
Mk VII, Tetrarch (A17)	177
Mk I, Cruiser Tank (A9)	125
Mk II, Cruiser Tank (A10)	175
Mk III, Cruiser Tank (A13)	65
Mk IV, Cruiser Tank (A13 Mk II)	665
Mk V, Cruiser Tank Covenanter (A13 Mk III)	1700
Mk VI, Cruiser Tank Crusader (A15)	5300
Mk VII, Cruiser Tank Cavalier (A24)	500
Mk VIII, Cruiser Tank Centaur (A27L)	950
Mk VIII, Cruiser Tank Cromwell (A27M)	3066
Mk VIII, Cruiser Tank Challenger (A30)	200
Comet I Cruiser Tank (A34)	1186
Centurion I Cruiser Tank (A41)	6
Total	32,885

APPENDICES

US AFV production

Although the USA's pre-war tank-building facilities were limited, the country's huge automotive industry was able to quickly adapt. US factories were more modern than their European counterparts and American mass-production methods were well suited to wartime requirements.

Model	1940	1941	1942	1943	1944	1945	Total
Stuart Light Tank, M1	34	–	–	–	–	–	34
Stuart Light Tank, M2	325	40	10	–	–	–	375
Stuart Light Tank, M3	–	2551	7839	3469	–	–	13,859
Stuart Light Tank, M5 and M8 HMC	–	–	2825	4063	1963	–	8851
Light Tank, M22 Locust	–	–	–	680	150	–	830
Light Tank, M24 Chaffee	–	–	–	–	1930	2801	4731
Gun Motor Carriage, M18 Hellcat	–	–	–	812	1695	–	2507
Medium Tank, M2A1	6	88	–	–	–	–	94
Medium Tank, M3 Lee/Grant	–	1342	4916	–	–	–	6258
Medium Tank, M4 Sherman (75)	–	–	8017	21,231	3504	651	33,403
Medium Tank, M4 Sherman (76)	–	–	–	–	7135	3748	10,883
Medium Tank, M4 Sherman (105)	–	–	–	–	2286	2394	4680
Gun Motor Carriage, M10	–	–	639	6067	–	–	6706
Gun Motor Carriage, M36	–	–	–	–	1400	924	2324
Howitzer Motor Carriage, M7	–	–	2028	786	1164	338	4316
Gun Motor Carriage, M12	–	–	60	40	–	–	100
Cargo Carrier, M30	–	–	60	40	–	–	100
Heavy Tank, M26 Pershing	–	–	–	–	40	2162	2202

Allied/Axis AFV losses by battle/campaign

Battle	Date	Allied	Axis
Arras Counterattack	21 May 1940	57	30
Sidi Barrani	9–12 December 1940	–	73
Bardia/First Battle of Tobruk	3–22 January 1941	–	210
Beda Fomm	5–7 February 1941	less than 12	100
Operation Crusader	18 November – 7 December 1941	278	300
First Battle of El Alamein	1–27 July 1941	193	100
Dieppe Raid	19 August 1942	28	–
Alam el Halfa	30 August – 2 September 1942	68	49
Second Battle of El Alamein	23 October – 4 November 1942	500	500
Medinine	6 March 1943	–	50
Falaise Pocket	13–21 August 1944	n/k	724
Ardennes Offensive	16 December 1944 – 16 January 1945	800	800

Tank armaments

British

Size	Type of Gun	Calibre	Type of Round	Muzzle Velocity f.p.s.	Projectile Weight lbs/oz	Penetration at 30° range in yards						Effective Range in yards		
						100	500	1000	1500	2000	2500	HE	Smoke	Shot
2pdr 40mm 1.575"	OQF 2pr Gun Mks IX and X	50	AP APCBC SV	2800 2600 4200	2–6 2–11 1–0	– – –	– 53 88	40 49 72	– 44 60	– 40 48	– – –	– 2000 –	– – –	– – –
3pdr 47mm 1.85"	OQF 3pr 2cwt Gun Mk II	40	APHE	1840	3–4	–	–	25	–	–	–	–	–	2000
6pdr 57mm 2.24"	OQF 6pr 7cwt Gun Mk 3 or 5	50	APCBC APDS	2630 4000	7–2	– –	87 131	80 117	73 103	67 90	– –	5500 –	– –	2000 –
75mm 2.95"	OQF 75mm Gun Mk V	36.5	APC APCBC	2030 2650	14–4	– –	68 103	61 94	54 86	47 78	– –	10000 –	4500 –	2000 –
76.2mm 3"	OQF 77mm Gun Mk 2	50	APCBC APDS	2575 3675	17 7–11	– –	120 182	110 165	100 148	90 130	– –	7000 –	2000 –	2000 1500
17pdr 76.2mm 3"	OQF 17pdr Gun Mk 2	55	APCBC APDS	2900 3950	17 7–11	– –	125 187	118 170	110 153	98 135	– –	12000 –	– –	2500 1800
95mm 3.74"	OQF 95mm Tank Howitzer Mk I	20	HEAT HE	1650 1050	14–13 25	110 –	110 –	110 –	110 –	110 –	110 –	6000	3400	–

United States

Size	Type of Gun	Calibre	Type of Round	Muzzle Velocity f.p.s.	Projectile Weight lbs	Penetration at 30° range in yards						Effective Range in yards		
						100	500	1000	1500	2000	2500	HE	Smoke	Shot
37mm 1.46"	37mm M6 Gun	57	APC	2900	1.92	–	46	42	40	37	–	3000	–	1000
75mm 2.95"	75mm M2 Gun 75mm M3 Gun	31 40	APCBC APC	1930 2030	14.4 14.96	– –	– 70	62 59	48 55	40 50	– –	13300 14000	1500 –	3500 –
76.2mm 3"	76mm Gun M1A1 and M1A2	55	APCBC HVAP	2600 3400	15.44 9.4	– –	94 158	89 134	81 117	76 99	– –	14200	2000	3500
90mm	90mm M3 Gun	53	APCBC HVAP	2650 3350	24.1 16.8	– –	126 221	120 200	114 177	105 154	– –	19600	–	3500
105mm	105mm M4 Howitzer	25	HEAT	1250	29.2	100	100	100	100	100	100	12000	4800	–

KEY:

AP – Armour-Piercing
APC – Armour-Piercing Capped
APCBC – Armour-Piercing Capped, Ballistic Capped
APDS – Armour-Piercing, Discarding Sabot
APHE – Armour-Piercing High Explosive
HE – High Explosive
HEAT – High Explosive Anti-Tank
HVAP – High Velocity Armour Piercing
SV – Super Velocity

APPENDICES

Report by 9 RTR: Aspects of close support to infantry in Forest Fighting

AUTHOR NOTE: After the actions in the Reichswald in February 1945 34 Armoured Brigade commander asked 9 RTR to prepare a report on their experiences and the lessons they had learnt during those actions. This report, a summary of which is reproduced below, was used as the basis for a Brigade Conference on the matter of forest fighting.

General

1. An operation fought by tanks and infantry in close co-operation in forest country should not be looked upon as an entirely different type of warfare to an operation fought in normal European country. The same principles and rules apply, though they must, in many cases, be adapted to suit the unusual conditions of limited visibility and restricted manoeuvre. These two factors, as well as imposing many restrictions and difficulties on the actual fighting troops, make it extremely difficult for a commander to influence the battle, once he has launched his troops into an attack.

Training

2. For an operation in this type of country to be successfully conducted, it is imperative that co-operative training in forest country be carried out by the troops taking part. This Regiment had approximately one week's training in forest country prior to the action in the Reichswald. This was the absolute minimum required, and when possible a period of a fortnight should be made available.

Types of Forest

3. This Regiment has now fought in woods of many varieties. As a result of this experience it is considered that it is practicable for tanks to support infantry in forests where the trees are more than 12 feet high and 3 feet apart. It is not practicable, however, for tanks to support infantry in young plantations where visibility is nil. In such plantations it is impossible for tanks to keep in touch with the infantry, except in the rides, and tanks are unable to defend themselves against bazooka teams.
4. A coniferous forest presents less difficulty to the passage of tanks than does a deciduous forest. A Churchill tank will knock down a coniferous tree of 2 feet diameter, and a "Honey" a tree of 1 foot diameter. The tree is normally broken off at the base, though sometimes, and more particularly by 'Honeys' it is uprooted. The reason for this is that the 'Honey' tank tends to ride up the tree before pushing it over, while the 'Churchill' with a flat forward plate produces a more horizontal push. A deciduous tree is invariably uprooted, and the type of soil will have considerable effect on the size of the tree that can be pushed over. In the Reichswald it was found that a Churchill tank would push over a beech tree of 9 inches diameter but failed to push over a tree of 1 foot diameter. Crashing through trees at high speed is not feasible, as the shock of the impact breaks off the tops of the trees which fall on the commander's head.

Co-operation with Infantry

5. In view of the blindness of tanks in forests, it is essential that in all advances, both by day and night, deployed infantry should precede the tanks, irrespective of whether the tanks are advancing down the rides or through the forest. In order to avoid falling trees hitting the infantry, they must be approximately 30 yards ahead of the tanks. To ensure this, advances at night must be by bounds and movement light must be used. The infantry must carry some easily seen mark on their backs, and infantry and tank commanders must be close together. In one case, infantry carrying white mugs on their backs advanced 80 yards at a time, signalling back with red lights to the tanks behind them each time they halted. This system worked satisfactorily. Flank as well as frontal protection must be provided.
6. At night great difficulty was experienced by troop commanders in recognizing the company which they were supporting. It is recommended that each company should wear some distinguishing mark, which would render them easily recognizable.

Tank Formations

7. The value of tanks in forest country is more moral than material. It is, therefore, essential that the element of surprise should be maintained as long as possible. To do this, formations and tactics must be varied as much as conditions allow, always remembering that close contact with the infantry must be maintained and that tanks must be in visual contact. As a generalization it is reasonable to say that at night tanks within a troop should move in line ahead in close forest, and can only deploy in fairly open forest in daylight. Squadrons have, however, advanced successfully four troops up through the forest at night.

The following formations in the attack were adopted and proved satisfactory by one squadron:

(a) Four troops up; tanks in line ahead, advanced through the trees.
(b) A track used as a Centre Line with one troop deployed on either side. Two other troops followed after 10 minutes in same formation on a parallel ride. Deployment only possible in daylight.
(c) Two troops up; deployed in the forest, using two tracks as Centre Lines; one troop following close behind and another troop in reserve. Deployment only possible in daylight.
8. Although it is possible for tanks to deploy into and move through a forest it must be remembered that moving through this type of country imposes very severe strain on tanks and therefore, if contact is unlikely, tanks should move along tracks and rides, always ensuring that an infantry screen to front and flanks is maintained, and that on straight rides crests and crossings are carefully reconnoitred before being crossed.

Communications & Control

9(a) The squadron commander should travel well up behind his leading troops.
(b) The Recce Officer must travel in a Churchill tank.
(c) The No. 19 set gives adequate communications within the Regiment.
(d) Close liaison between flanking sub-units, units etc. must be maintained, or there is great danger of firing into one's own troops.

Artillery

10(a) SP Anti-tank artillery must be maintained well forward and be able to get into position quickly on an objective being gained.
(b) An FOO travelling with squadrons is most valuable and should always be provided when possible.

Night Leaguers

11. It is essential that as soon as an objective has been reached at night, tanks be withdrawn at least 20 yards from the foremost infantry positions and placed where they are adequately protected by the infantry. If possible it is highly desirable that tanks should be rallied by squadrons near the infantry battalion HQ (one squadron of tanks in support of an infantry battalion). The tanks must also have their own guards, armed with Sten guns at the ready. No movement of any kind within company positions should be allowed, and any person walking about must be shot without being challenged. In the Reichswald a German was able to penetrate a company position, and, on being challenged by a tank commander from his turret, shot him through the head. On another occasion a tank was hit by a bazooka at night while supposedly protected by the infantry dispositions.

Firing

12. During the night advances Besas were loaded with belts cleared of tracer and fired into the tops of the trees above the heads of the advancing infantry. It is considered that this has a useful horrific value.
During the attack it is of definite value to maintain Besa fire into the forest even when no more enemy are visible as it is found that those moving back out of sight become casualties.
13. In order to avoid inflicting casualties on one's own infantry, HE should normally be fired only in clearings.
14. The consumption of Besa ammunition is much in excess of all other types.

Recovery

15. ARVs should travel where they can give early assistance to bogged tanks. A liberal support of bulldozers well forward is most essential to clear trees off tracks and enable essential wheeled vehicles to go forward.

Supplies

16. Getting supplies forward presented great difficulty. The following methods were tried:-
(a) Churchills towing sledges. Sledges were too heavy to tow over heavy ground. Tanks could only tow in bottom gear with difficulty and eventually got bogged, but the sledges with wheels were much more satisfactory.
(b) RHQ tanks, carrying supplies forward. A certain but slow and laborious method. As these tanks obviously cannot be made available for this duty for long periods, it can only be looked upon as an emergency measure.
(c) Using two RHQ tanks (2IC and HQ Troop Commander's tanks) to tow forward two 3-tonners each. This is a feasible method and the majority of the supplies were taken forward in the Reichswald by this method.
(d) Recce Troop tanks used to carry supplies forward. This is the quickest and most efficient method, but special racks should be fitted to the tanks to increase their carrying capacity. In the Reichswald operation Recce Troop were unfortunately frozen by Traffic Control authorities while employed on this duty.

Bibliography

Forty, George. *US Army Handbook, 1939–1945.* Stroud, UK: Sutton Publishing, 2003.

Macksey, Kenneth. *Tank versus Tank.* London: Grub Street, 1999.

Man, John. *The Penguin Atlas of D-Day and the Normandy Campaign.* London: Viking, 1994.

Perret, Bryan. *Iron Fist, Classic Armoured Warfare Case Studies.* London: Brockhampton Press, 1999.

Porter, David. *'Armour in Battle'* articles in *Miniature Wargames Magazine.* Issues 176, 177, 178 (January-March 1998), 187, 188 (December 1998 – January 1999) and 200 (January 2000).

Websites

http://derela.republika.pl/armcarpl.htm
An excellent English language website covering Polish AFVs (and armoured trains) from 1918 to 1939.

http://france1940.free.fr/en_index.html#AdA
A treasure house of technical information on French AFVs of 1939/40.

http://www.wwiivehicles.com/default.asp
A very useful website for illustrations and technical data on a wide range of AFVs of all nationalities.

http://www.btinternet.com/~ian.a.paterson/main.htm
A highly detailed account of 7th Armoured Division from its formation in 1940 to the end of the war in Europe.

http://www.royaltankregiment.com/9th_RTR/TT/CONTENTS.HTM
This covers the war service of 9 RTR from its formation in 1940 to VE Day. Extracts from the unit's war diary and veterans' accounts are combined to give a remarkable account of life in a wartime tank regiment.

http://www.royaltankregiment.com/9th_RTR/tech/reichswald/Reichswald%20Report.htm
A report by 34 Armoured Brigade on operations in the Reichswald, February 1945. This provides an invaluable insight into some of the problems facing Allied armoured operations in the winter of 1944–45.

http://afvdb.50megs.com/
A highly detailed collection of technical data on almost all US AFVs.

http://www.onwar.com/tanks/index.htm
A good collection of specifications of British and US tanks.

Index

Page numbers in *italics* refer to illustrations.

4th Indian Division 55, 57
4TP light tank 12
6th Australian Division 57, 58, 60
7TP light tank *10*, 11, 12, 14, *18*, *19*
10TP fast medium tank 12

Aachen 138
Abbeville 35
Achilles tank destroyer *99*
AEC Armoured Car Mark III *93*, *164*
Aisne, River 39, 41
Alam el Halfa, Egypt 69, 70, 71
Albania 62
Algeria 72, 74
AMC Schneider P16 halftrack *41*
Amiens, France 7, 38–9
Antwerp 130, 138–9, 151
Anzio 88–90
APDS (armour-piercing, discarding sabot) ammunition 120–1
appliqué armour 138, *150*
Archer 17pdr self-propelled gun, Valentine Mark I 121, *124*
Ardennes
 1940 32
 1944 137, 138–9, 141–6, 148–51, 153, 154, 174
armaments, tank 183
Armoured Bulldozer 115
Arnhem 132, 135
Arnim, Hans-Jürgen von 73, 74
Arras 47–8, *49*, 63
Arromanches 125, *126*
Artillerie d'Assaut 25
Atlantic Wall 103
Auchinleck, General Claude 66, 68
Austin 10 Light Utility Truck *83*
Australia, North Africa 57, 58, 60
Austria 168
autotransports 9

Badoglio, Marshal Pietro 85
Bailey Bridge 163
Balbo, Air Marshal Italo 52
Balkans 139
Bardia 57–8
Barham, HMS 58
Bastogne 141, 144, 145, 149, 150, 151
Bayeux 124
Beda Fomm, Libya 61, 64, 65

Bedford
 3-ton Fuel Tanker *135*
 MWD GS 15cwt Truck 4x2 *135*
 OYD GS 3-Ton Truck *45*
Belgium
 1re Division de Cavalerie 25
 1re Division des Chasseurs Ardennais 24
 2e Division de Cavalerie 24
 France and 24
 German invasion 22, 24, 31, 32, 34–5, 122
 strength 22, 24
Benghazi 59
Bir Hacheim 68
Bishop 25pdr self-propelled gun 81
bocage 127, *128*, 129
Boulogne 48
Brandenberger, General Erich 141–2, 146
Bremen 165
Bren Gun Carrier *47*
British Expeditionary Force (BEF) 31, 37, *43*, 44, *46*, 47–8, *49*
Brochów, Poland 14
Bron Pancerna (Armoured Forces) 11, 16
Brussels 130, 139, 151
Buerat, Libya 73
Bzura, River 14

C15TA Armoured Truck *155*
Caen *122*, 124, 129–30
Cairo 52, 54, 65
Calais 48
camouflage *15*, *56*, *62*, *68*, *72*
Canada
 First Canadian Army 155
 I Canadian Corps 92, 98
 4th Canadian Armoured Division 180
 5th Canadian Armoured Division 180
 Germany, invasion of 155
 Italy *76*, 91–2, *98*
 losses 92
 Normandy landings *116*, *117*, *123*, 124, *129*
Carden-Loyd tankettes 11, 17
Centaur Bulldozer 115–16
Centurion I Cruiser Tank 181

Chaffee, Brigadier-General Adna 172
Challenger tank 121, *161*, 181
Char 2C 25, 28, *37*, 181
Char B heavy tank 26, 34–5, 39
Char B1 bis heavy tank *30*, *31*, *36*, *38*, *39*, *40*, 181, *181*
Char D medium tank 26, *181*
Char D2 medium tank 35, *181*
Chars de manoeuvre d'ensemble 26
Cherbourg 126, 130, 131
Chevrolet
 C60L *164*
 WA *60*
Christie suspension 8–9, 12
Churchill, Winston 77, 124
Churchill Armoured Ramp Carrier (ARK) *109*, 111, 113
Churchill AVRE *102*, 107, *108*, *109*, 110–13, 162–3
Churchill Crocodile Flamethrower *105*, 108, *109*, 110, 119, 154, 165
Churchill Mark III, Infantry Tank Mark IV *128*, 181
Churchill Mark IV, Infantry Tank Mark IV 73, *101*, 110, *120*, 181
Citroën-Kegresse P19 halftrack *42*
Clark, General Mark 85, 89–90
Cold War 169
Cologne 174
Combe, Lieutenant-Colonel John 61
Cracow 14
Cromwell Armoured Recovery Vehicle (ARV) *114*
Cromwell Mark IV *128*, *171*
Cromwell Mark VII *163*
Cruiser Tank Mark I, A9 Mark 1 *62*, 181
Cruiser Tank Mark II, A10 Mark 1A *53*, *54*, *61*, *62*, 181
Cruiser Tank Mark III, A13 *181*
Cruiser Tank Mark IV, A13 Mark II *46*, *54*, *55*, *61*, *62*, 181
Cruiser Tank Mark V, Covenanter A13 Mark III *181*
Cruiser Tank Mark VI, Crusader Mark I 9, *64*, *65*, 181
Cruiser Tank Mark VI, Crusader Mark III *70*, 81
Cruiser Tank Mark VII, Cavalier A24 *181*

INDEX

Cruiser Tank Mark VIII, Centaur A27L *181*
Cruiser Tank Mark VIII, Cromwell A27M 9, *181*
Cunningham, General Alan 66
Cyrenaica 52
Czechoslovakia 14, 168

D-Day *see* Normandy landings
Daimler
  Armoured Car *69*, 73, *91, 93, 164*
  Scout Car Mark IA (Dingo) *91, 93*
  Scout Car Mark II (Dingo) *151*
Danube, River *159*
de Gaulle, General Charles 26, 31
Derna, Libya 65
Dessau, Germany 174–5
Dieppe Raid 103, 104
Dietrich, SS General Josef 'Sepp' 141, 142–3, 151
Dinant, Belgium 33, 34, 149
Divisions Cuirassée de Réserve (DCR)
  1re 26, *30, 31,* 33–4, *42*
  2e 26, *33,* 34, *36,* 41
  3e *33,* 34, *38,* 39, 41
  4e 26, *27, 33, 39, 40,* 41
Divisions Légère de Cavalerie (DLC) 28, *34*
  1re *34, 42*
  2e *34,* 39
  3e *34*
  4e *34*
  5e *34,* 39
Divisions Légère Mécanique (DLM) 28
  1re *8,* 26, *34, 36, 38, 39*
  2e 32, *34*
  3e *27, 29,* 32, *34, 36,* 47, *49*
  7e 41
Dodge
  WC51 Weapons Carrier *96, 147*
  WC54 4x4 Ambulance *94*
  WC58 Command Reconaissance Radio Car *96*
Dortmund-Ems Canal 163
DUKW *85, 159*
Dunkirk 47, 48, 130
Dyle, River 32

East Prussia 14
Egypt 44, 52, 65, 66
Eindhoven 132, 135
Eisenhower, General Dwight 105
El Agheila, Libya 73
El Alamein, Egypt 51, 68, 70–2
Elbe, River 112, 113, 166
English Channel 37, 38, 40, 48

'Enigma' codes 65
Eritrea 55
Estienne, General Jean-Baptiste 25
Ethiopia 55

Falaise Pocket 123, 130
FCM-36 light tank *37, 181*
Fiat-Ansaldo
  M11/39 medium tank 53, 55, *57,* 58
  M13/40 medium tank 58, 59–60, 61–2
Focke-Wulf Fw 190 150
Ford
  C11 ADF Car, Heavy Utility, 4x2 *82*
Ford/Marmon-Herrington Armoured Car *24*
  GPA Seagoing Jeep (Seep) *86*
  T17E1 armoured car *88*
  WOA *170*
  WOA2 *170*
forest fighting 184–5
France
  First Army *125, 126,* 168
  1st Armoured Division *125, 126,* 180
  2nd Armoured Division *123, 125,* 130, 180
  5th Armoured Division 180
  AFV production *181*
  armour, evolution of 25–6, 28–9
  battle for 31–5, 37–41, 44, 47–8, *49*
  Belgium and 24
  German invasion 28–9, 122
  Germany, invasion of 168
  insignia *32*
  losses 32
  Normandy landings 103, 116, 120–1, 122–6, *127–8,* 129–32
  North Africa 68, 74
  Poland 12, *15*
  rearmament 12
  strength 21, 28–9, 32, *33–4,* 37, 41
  surrender 41, 52
  World War I 7, 8, 25, 43
  see also Divisions
Fuller, J.F.C. 7

Gazala, Libya 67, 69
Germany
  1st Light Division 16
  2nd Marine Division 165
  4th Light Division 14
  5th Light Division 64

Fifth Panzer Army 130, 141, 143–4, 146
Sixth SS Panzer Army 141, 142, 151
Seventh Army 130, 142, 146
Tenth Army 89, 90
Fourteenth Army 89, 90, 97
XIV Panzer Corps 78
15th *Panzergrenadier* Division 78
XIX Panzer Corps 34, 48
20th SS Training Division 165
XLI Panzer Corps 34
Allied invasion of *152,* 153–6, 160–3, 165–9, 174–5
Ardennes Offensive 138–9, 141–6, 148–51, 174
armour, developments in 29–30
Army Group A 32, 41
Army Group B 32, 40–1, 122, 161
Army Group G 168
Army Group Ostmark 168–9
Deutsche Afrikakorps (DAK) 66, 67–8, 69, 70, 72, 74
France, battle for 31–5, 37–41, 44, 47–8
France and the Low Countries 21, 22, 23, 24, 28–9, 122
fuel shortages 71, 148, 150, 151
Italy 77, 85–6, 88–92, 95, 97–8
losses 14, 16, 32, 66, 72, 130, 150, 161, 175
Normandy landings 103, 122–5, 126, 129, 130
North Africa 64–9, 70–5
Operation *Market Garden* 134–5
Panzergruppe Afrika 66
Panzergruppe West 123
Poland, invasion 14, *15,* 16
Sicily 78–9, 80
strength 12–13, 29–30, 32, 72–3, 78, 97, 141, 156, 161
see also Panzer Divisions
Gothic Line 90, 91, 92, 94
Grant Canal Defence Light (CDL) *115*
Graziani, Marshal Rodolfo 52, 53–4, 57, 61, 97
Greece 62
Guderian, General Heinz 12
  France, battle for 35, 41
Gustav Line 86, 89
Guzzoni, General Alfredo 78

Halfaya Pass 63, 65
Hamburg 163, 165, 167
Hamilcar glider 122

188

INDEX

Hitler, Adolf
  Ardennes Offensive 137, 138–9, 145
  France 48
  Italy 86
  Normandy landings 123, 124
  North Africa 72, 74
  Versailles Treaty 12
Hitler Youth 161, 163, 165, 166
Hobart, Major-General Percy 52, 104, *114*
Hotchkiss
  H-35 light tank *8*, *25*, 26, *27*, 32, 35, *38*, *181*
  H-39 26, *27*, 32, *36*
Houffalize, Belgium 148, 150
Humber
  Armoured Car Mark II *73*
  Car, Heavy Utility 4x4 *82*
  Snipe Light Utility Truck *83*

infantry, close support to 184–5
insignia
  American *100*, *133*, 176–8
  Australian *57*
  British *44*, *62*, *68*, *81*, *104*, *128*, 178–9
  Canadian 180
  French *32*, *37*, *38*, *125*, 180
  Polish 180
  South African 180
Italy
  Sixth Army 78
  Allied invasion *76*, 77, 85–6, 88–92, 94, *95–6*, 97–8, *99–101*, 168
  Bambini Armoured Brigade 59, 60, 61–2
  Ligurian Army 97
  Livorno Division 78
  losses 57, 60, 62
  North Africa 52–5, 57–62, 66–7, 68, 70–1, 74
  Sicily 78–9
  strength 52, 54, 58, 66–7, 71, 97

Jagdpanther 156
Jagdtiger 161
Junkers Ju 87 150

Kangaroos *109*, 116, *117*, 163
Kasserine Pass 74
Kesselring, Field Marshal Albert 86, 88, 97
Kiel 167
Konev, Marshal Ivan 168
Kriegsmarine 12

L3 light tank 53, 58
Laffly S15T tractor *41*
Landing Vehicle Tracked (LVT-2) Buffalo II *115*
Landing Vehicle Tracked (LVT-4) Buffalo IV 97, *98*, 114–15
Leclerc, General Philippe 125
Leese, Sir Oliver 82
Lemelsen, General Joachim 90
Libya 57, 62
Liège 141
Lorraine 38L (VBCP) armoured personnel carrier *42*
Ludendorff Bridge, Remagen 157, 160
Luftwaffe 12
  Ardennes Offensive 149–50
  Belgium 32
  France 33
  Germany, defence of 160, 165
  North Africa 66
  Poland 14
Lutz, General Oswald 12
Luxembourg 142

M2 halftrack 82, *89*, 93
M2A1 halftrack *86*
M2A4 Light Tank *172*
M3 75mm Gun Motor Carriage (GMC) 173
M3 Grant/Lee medium tank *50*, 64, 68, 70, *75*, 172, *182*
M3 halftrack 81, *84*, *93*, *143*, *146*, *163*, *167*, *173*
M3 Stuart I Light Tank 65, 66, *67*, *73*, 100, *183*
M3A1 halftrack *134*
M3A1 Scout Car *144*
M3A1 Stuart light tank *74*, *84*
M3A3 Stuart light tank *125*, *151*
M4 81mm Motor Mortar Carriage (MMC) *82*, *134*
M4 Composite Medium Tank *138*
M4 Dozer *79*
M4 High-Speed Tractor *146*
M4 Sherman Crocodile Flamethrower *119*
M4 Sherman Medium Tank 64, *69*, 70, *72*, *76*, *84*, *87*, *89*, *120*, 149, 174, *182*
M4A1 Sherman Medium Tank *79*, *80*
M4A2 Sherman Medium Tank *114*, *122*, *123*
M4A3 Sherman Medium Tank *88*, *121*, *157*, *169*
M4A3E8 Sherman Medium Tank *139*, *165*, *171*

M5 High-Speed Tractor *94*, *147*
M5 Light Tank *150*
M5A1 Light Tank *100*, *121*
M6 37mm Gun Motor Carriage (GMC) *173*
M7 Priest 105mm Howitzer Motor Carriage (HMC) 70, *80*, *81*, *84*, 116, *131*, *142*, *167*, *182*
M8 75mm Howitzer Motor Carriage (HMC) *121*
M8 Armoured Car *87*, *124*, *125*, *145*, *162*, *163*, *168*
M10 3in Gun Motor Carriage (GMC) *95*, *99*, *121*, *130*, *173*, *182*
M12 155mm Gun Motor Carriage (GMC) *131*, *182*
M18 76mm Gun Motor Carriage (GMC) 'Hellcat' *148*, *182*
M19 Gun Motor Carriage (GMC) *160*
M20 Armoured Utility Car *126*, *133*
M22 Locust Light Tank *154*, *182*
M24 Chaffee Light Tank *100–1*, *136*, *141*, *158*, *166*, *169*, *182*
M26 Pershing Heavy Tank 145, *152*, *157*, 160, *166*, *170*, *175*, *182*
  combat trials 174–5
M26A1E2 Pershing Heavy Tank 174
M29 Weasel *168*
M30 Cargo Carrier *182*
M32 Tank Recovery Vehicle *132*
M36 90mm Gun Motor Carriage (GMC) *95*, *130*, *145*, *146*, *148*, *182*
M39 *Pantserwagen* 23
M40 155mm Gun Motor Carriage (GMC) *157*
M1917 tank *172*
Maas, River 154–5
MacArthur, General Douglas 172
Maduri, Ssgt Joe 174–5
Maginot Line 21, 24, 31
Maletti, General Pietro 57
Manstein General Erich von 32
Manteuffel, General Hasso von 141
Mareth Line 74
Mark VI light tank *43*, 44, *52*
Mark VIB light tank 44, 45, 47, 53
Mark VIC light tank *45*
Mark VIII heavy tank (US) 172
Marmon-Herrington Armoured Car Mark II *59*, 60
Marne, River 41
Matilda I infantry tank *43*, 44, 47, 48, 49, *181*

INDEX

Matilda II infantry tank *43*, 44, 47, *48, 49, 55, 56*, 57, 58, *181*
Matilda Scorpion flail tank 70
Mechanized Brigades (OMs), Polish 12, 14, 19
Mellenthin, Major-General Friedrich von 148–9
Mersa Matruh 52, 54, 68
Messerschmitt Bf 109 165
Meuse, River 32, 38, 139, 144, 146, 149, 151
Mine Exploder T1E3 (M1) 'Aunt Jemima' 118, *119*
Model, Field Marshal Walter 138, 161
Monte Cassino 86
Montgomery, General Bernard
  Ardennes (1944) 146
  Germany, invasion of 160
  Italy 85
  Normandy landings 129, 130, 131–2
  North Africa 51, 70, 71, 74
  Operation *Market Garden* 132
  Sicily 78
Montherme, France 33, 34
Morocco 72
Morris
  C8 Mark III Artillery Tractor 4x4 *156*
  CS8 GS 15cwt Truck *46*
  CS9 armoured car 44, *46, 59*, 61
Mulberry Harbours 124, 125
Mussolini, Benito
  North Africa 52, 57, 62, 72

Naples 85
Netherlands
  German invasion 23, 31, 32, 122
  Operation *Market Garden* 132, 134–5, 138, 160
  strength 22–3
New Zealand
  2nd New Zealand Division 88
  Italy 88
  North Africa 66
Nijmegen 132, 135, 154
Normandy landings 103, 116, *117*, 120–1, 122–6, *127*–8, 129–32
North Africa *50*, 51–6, 57–75

O'Connor, General Richard 52, 54, 57, 58, 59, 60, 62, 65
Operations
  Battleaxe 63, 65, 66
  Blackcock 154
  Brevity 65, 66
  Charnwood 129

Cobra 130, 133
Compass 54–5, 57
Crusader 66
Dynamo 48
Epsom 129
Fall Gelb 21, 32
Goodwood 129–30
Grenade 156
Market Garden 132, 134–5, 138, 160
Plunder 160
Torch 72–3
Totalize 116
Varsity 160
Veritable 154
Wacht am Rhein 138–9, 141–6, 148–51
organization 9
  First US Army *138*
  1re Division Légère Mécanique (DLM) *26*
  1re Divisions Cuirassée de Réserve (DCR) *30, 31*
  1st Canadian Armoured Carrier Regiment *116, 117*
  1st US Armored Division *74, 90*
  2nd French Armoured Division *123*
  2nd US Armored Division *78, 84*
  III US Corps *143*
  Third US Army *138*
  4th Armoured Brigade (UK) *129*
  6th South African Armoured Division *99*
  7th Armoured Division (UK) *64, 67, 127*
  VII US Corps *140*
  Eighth Army (UK) *67, 70, 71, 90*
  VIII US Corps *143*
  10th Polish Mechanized Brigade *14*
  12th US Armored Division *156*
  XII US Corps *144*
  XVIII US Airborne Corps *140*
  27th Armoured Brigade (UK) *113*
  XXX British Corps *80*
  79th Armoured Division (UK) *104, 109*
  Belgian armoured forces *22*
  British armoured division *43*, 44
  Divisions Légère de Cavalerie *34*
  French armoured forced *33*–4
  Polish tank formations *13*
  US Fifth Army *89*
  US Seventh Army *78*
  US Armored Divisions *158, 173*
  US tank/tank destroyer battalions *130, 139, 145, 159, 173*

Warsaw Mechanized Brigade *19*
Ortona, Italy *76, 92*

Palermo, Sicily 80
Panhard 178 Armoured Car *30, 42*
Pantherturm 91
Panzer Divisions
  1st 34
  1st SS 122
  2nd 14, 34, 144
  3rd 32
  4th 14, 32
  5th 33–4
  6th 34
  7th 33–4, 47
  8th 34
  9th 123, 148
  10th 34
  12th SS 122
  15th 72
  16th 85
  21st 73–4, 124
  116th 160
  Hermann Göring 78
  Lehr 122
Panzerfaust 126, 138, 154, 156, 161, 175
Panzerjäger *Nashorn* 174
Panzerkampfwagen I 12, 29, 30, 32
Panzerkampfwagen II 12, 29, 30, 32
Panzerkampfwagen III 12–13, 29, 30, 32
Panzerkampfwagen IV 12–13, 30, 32, 48, 70, 89, *114*, 129
Panzerkampfwagen V Panther 150–1, 153, 174, 175
Panzerkampfwagen VI Tiger 70, 129, 150, 174, 153153
Panzerkampfwagen VI Tiger II 141, 175
Panzerkampfwagen 35(t) 13, 16, 29, 30
Panzerkampfwagen 38(t) 13, 29, 30, 47, 48
Panzerschreck 156
Paris 125, 130
Patton, General George *133*
  Ardennes (1944) 146, 150
  Desert Training Center 172
  Germany, invasion of 168
  Normandy landings 131–2
  Sicily 78
Phoney War *22*
Po, River 97
Poland
  1st Armoured Division 180

INDEX

1st Polish Parachute Brigade 132
Bron Pancerna (Armoured Forces) 11, 16
 German invasion 14, *15*, 16
 Italy *97*
 Normandy landings *122*
 rearmament 12
 Soviet invasion 16, 139
 strength 11, 12, *13*
Polski Fiat PF-618 Light Truck *18*
Pomerania 14
Porcinari, General Giulio 78
Puszcza Kampinoska, Poland 14, 16

Qattara Depression, Libya 71

Ram Kangaroo *109*, 116, *117*
Ram OP/Command Tank *127*
Remagen 157, 160
Renault
 ACG-1 Light Tank *22*
 AMR-33 light tank *28*, *181*
 AMR-35 light tank *181*
 Chenillette prototype *26*
 FT 25, 28
 FT-17 *6*, *12*, *15*, *40*, *172*, *181*
 G1 R heavy tank *181*
 R-35 tank 12, *20*, *26*, *27*, 28, *29*, 35, *181*
Rendulic, General Lothar 169
Rhine, River 112, *113*, 153, 157, 160
Ritchie, General Neil 66, 68
'Roer Triangle' 154–5, *156*
Rolls Royce armoured car 61
Rome 90
Rommel, Field Marshal Erwin
 France 47, *48*
 Normandy landings 103, 122–3, 124
 North Africa 51, 63, 64, 66–9, 70, 71–2, 74–5
Royal Air Force, North Africa 58
Royal Navy, Mediterranean 58
Royal Tank Corps 43
Ruhr 160–1
Rundstedt, Field Marshal Gerd von 41, 123, 138, 139

St Vith, Belgium 141, 143–4, 145, 150
Salerno 85, *87*
San Giorgio 58
Scheldt, River 130
Scout Carrier 44, *47*
SdKfz 8 halftrack *30*
Sedan, France 33, 34, *37*

Semovente 90/53 self-propelled gun 71
Sexton 25pdr self-propelled gun *126*
Sherman Beach Armoured Recovery Vehicle (BARV) *113*, 114
Sherman Crab Flail Tank *106–7*, *109*, 114, 118, *162*, *163*
Sherman Duplex Drive amphibious tank 107, *112*, *113*, 114, 123
Sherman Firefly *97*, 121, *129*, *149*, *155*, 161
Sicily 78–81, *82*, 83, *84*, 132
Sidi Barrani 52
Sidi Rezegh, Libya 66
Siegfried Line 154, 155
Silesia 14
Slovakia 14
Sollum, Egypt 65
Somme, River 38, 39
Somua S-35 12, 26, *29*, 32, 34, 35, *39*, *47*, *49*, *180*, *181*
South Africa 99
 6th South African Armoured Division 180
Soviet-German Non-Aggression Pact (1939) 16
Soviet Union
 Germany, invasion of 168
 Poland 16, 139
Speer, Albert 151
Staghound AA Armoured Car *90*
Staghound Mark I Armoured Car *88*, *93*
Stalin, Josef 16
Studebaker US6 2?-Ton Truck *147*
Stumme, General Georg 72
Sturmgeschütz III *30*, 70, *163*
Sturmtiger 156
Sudan 55
Sudetenland 14
Syracuse, Sicily 78

T13 tank destroyer *22*, 24
T15 light tank *22*, *25*
T15E1 Mine Resistant Vehicle *118*
T19 105mm Howitzer Motor Carriage (HMC) *81*
T-34 tank 9
tank armaments 183
Tetrarch ICS Light Tank Mark VII *122*, 154, *181*
Teutoburger Wald 163, 165
TK tankette *9*, 11, *12*, 14, *17*
TKS tankette 11, *12*, 14, 16, *17*, 19
Tobruk 57, 58–9, 65, 66, 67–8
Torgau, Germany 168

Truscott, General Lucian 90, *98*
Tunisia 72–3, 74, 75

Unic P107 halftrack *36*
United Kingdom
 First Army 73, 75
 1st Airborne Division 132
 1st Armoured Division 38–9, 44, *45*, *46*, *71*, 178
 1st Army Tank Brigade 47, *48*
 1st Infantry Division 88
 2nd Armoured Division *59*, 178
 2nd Infantry Division 45
 Second Army *124*, *126*, *127*, *128*–9, 132, *135*, *149*, *151*, 154–5, *156*, *163*–4, *170*
 3rd Infantry Division *112*, *123*
 4th Infantry Division 73
 6th Airborne Division *122*, 154
 6th Armoured Division 74, 179
 7th Armoured Division 179
  Germany, invasion of 154, *155*, *163*, *164*, 165, *171*
  Italy 85
  Normandy landings *127*, *128*, *129*, *135*
  North Africa 44, *53*, *54*, 55, *57*, 58, *59*–60, 65, 66, 71, 73
 Eighth Army
  Italy 85, *88*, *90*, 94, *97*, *98*
  North Africa *59*, 60, *63*, 65, 66, 68, 70, *71*, 73, 74, *75*
  Sicily 78, *81*, *82*–3
 8th Armoured Division 179
 VIII Corps *124*, *128*, 133
 9th Armoured Division 179
 10th Armoured Division *71*, 179
 11th Armoured Division 179
 XII Corps 154, *155*
 XIII Corps 57, *63*, *81*, 83
 15th (Scottish) Infantry Division *128*, *132*, 155
 Twenty-First Army Group
  Ardennes (1944) 146, *149*, *151*
  Germany, invasion of 155–6, 160, 161–2, *163*–4, *170*–1
  Normandy landings *124*, *126*, *127*, *128*, *129*, 130, *132*, *135*
 XXX Corps 66, 70, *82*, 83, *126*, *127*, *135*, *149*, *151*, 155, *156*, *164*
 42nd Armoured Division 179
 43rd Wessex Division 154, *155*
 51st Highland Division *37*, 155, *164*
 52nd Lowland Division 154, 165
 53rd Infantry Division 155, 165

INDEX

79th Armoured Division
103, 105–6, *106*, 107–8, *109–11*, *113*, *115*, *117*, 118, 119, 154, 160, 179
AFV production *181*
Ardennes (1944) 139, 146, 149
Combeforce 61–2
Experimental Mechanized Force 8, 43–4
France and the Low Countries 31, 37, 38–9, *43*, 44, *46*, *47–8*, *49*
Germany, invasion of 154–6, 162–3, *164*, 165–7, *170–1*
Guards Armoured Division 130, 134–5, 155, *156*, 179
Italy 85, 88–90, 91–2, *93*, 97–8, *99*, *101*
losses 47, 48, 62, 66, 69, 72, 94, 124, 130, 154, 163
Normandy landings *106*, 120–2, 123, 124, 125–6, *127–8*, 129–32
North Africa 72–3
Operation *Market Garden* 132, 134–5, 138, 160
pre-war 43–4
Sicily 78, *80*, 81, *82–3*
strength *43*, 44, 52, 54, 65, 66, *71*, 72, 78, 107, 155, 181
Western Desert Force 52, *53*, 54, *55*, *56*, 57, *58–9*
World War I 7, 8, 43
United States
First Army *121*, 130–1, 132, *134*, 139, *142*, 146, 150, *157*, 160, 161, *166–7*, *171*
First US Army Group 125
1st Armored Division 73, 74, *75*, 89, *94*, 100–1, 176
1st Infantry Division *130*
II Corps *94*
2nd Armored Division 78, *79*, 80–1, *82*, *84–5*, *121*, *131*, 132, *134*, *142*, 168–9, *171*, 176
2nd Infantry Division 139, 142
III Corps *142*, 143, 144, *146*, *147*
3rd Armored Division *157*, 166–7, *170*, 174, 176
3rd Infantry Division 88, 89, *95*
Third Army *123*, *125*, 133, 138, 139, *141*, *143*, 144, 146, *146–8*, 150, *157–8*, *162*, 168
IV Corps 87, 94, 100–1
4th Armored Division 143, 144, 146, *147*, 176
4th Infantry Division *131*
Fifth Army 85, 86, *87*, *88*, 89, *91*, 94–6, 100–1
5th Armored Division 176
V Corps *121*, 123, 125, *131*, *132*, *134*, 142
Sixth Army Group 168
6th Armored Division *91*, 138, *139*, 177
VI Corps 86, 88, 89, 90, *95*
Seventh Army 78, *79*, *81*, *82*, 85, 86, 87
7th Armored Division 177
VII Corps *130*, *131*
8th Armored Division 160, 177
VIII Corps 143, 146–7, *148*
Ninth Army 119, 138, 155, 156, 160, 161, 167–8
9th Armored Division 151, *157*, *158*, 160, *162*, 174, 177
10th Armored Division 144, 177
11th Armored Division 177
12th Armored Division *156*, 177
Twelfth US Army Group
Ardennes (1944) *138*, 139, *141*, 142–3, 144, 146–8, 150
Germany, invasion of *157–8*, 160, *162*, 166–9, *170–1*
Normandy landings 130, 131, *133*
XII Corps 144
13th Armored Division 178
XIII Corps *91*
14th Armored Division 178
16th Armored Division 178
20th Armored Division 178
29th Infantry Division *168*
30th Infantry Division 160
71st Infantry Division *168*
82nd Airborne Division 132, *141*, 144, *148*
87th Infantry Division 146, *147*
88th Infantry Division *94*
99th Infantry Division 139, 142
101st Airborne Division 132, 144, *148*
106th Infantry Division 139
AFV production *182*
Ardennes (1944) 138–9, *140*, 141–4, 145–6, 174
armour, development of 172–3
Germany, invasion of *152*, 155, 156, 160–1, *166*, 167–9, *171*, 174–5
Italy 85–6, 88–90, 91–2, 94, 95–6, 97–8, 100–1
losses 73, 74, 89, 123, 143, 150, 151, 174
Normandy landings 107, 120, *121*, 123, 125–6, 130–2
North Africa *50*, 51–5, 57–75, 74–5
Operation *Market Garden* 132, 135
Sicily 78–81, *82*, 83, *84*
strength 78, 139, 141, 150, *158–9*, 173
World War I 8, 172
Universal Carrier *43*, 44, 47, *58*, *96*, *101*, 132

Valentine Mark II, Infantry Tank Mark III *63*, 81, *181*
Valiant, HMS 58, 85
Versailles, Treaty of (1919) 12
Vickers
6-ton Mark E tank *10*, 11, *12*, 18
Carden-loyd tankettes 11, 17
Medium 8, 43
Utility Tractor *23*
Vienna 168
Vietinghoff, General Heinrich von 90, 97–8
Villers-Bocage 129
Volkssturm 161, 166

Waal, River 154–5
Warsaw 14
Warsaw Mechanized Brigade *13*, 14, *19*
Warspite, HMS 58, 85
Wasp Mark IIC Flamethrower *97*, *98*, 108, 165, 166
Wavell, General Archibald 54, 57, 65, 66
Weygand, General Maxime 40
Willys MB Jeep *87*, *145*
World War I 7, 8, 25, 43, 172
wz.28 armoured halftrack 15
wz.29 armoured car *12*
wz.34 armoured car *12*, 15, 17